# Racecar Technology Level Two

## To The Race Car – How To Setup For Competition

**Bob Bolles**

**CRD Publishing**

**CRD Publications**

Published by Chassis Research and Development Corporation, aka/Chassis R&D

P.O. Box 730542, Ormond Beach, FL 32173-730542. USA

Email: chassisrd@aol.com

# Race Car Technology – Level Two

Copyright © 2019 by Chassis Research and Development Corporation

All photos by Bob Bolles unless otherwise noted.

All rights reserved. No part of this book/publication may be reproduced, scanned, or distributed in any printed or electronic form without permission. Please do not participate in or encourage piracy of copyrighted materials in violation of the author's rights. Purchase only authorized editions.

CRD Publications is a trademark of Chassis Research and Dev. Corp.

PRINTING HISTORY

First CRD print edition / June 2019


ISBN: 978-1-7324884-4-1

CREATED/PRINTED IN THE UNITED STATES OF AMERICA


Cover design by: LS Bulow, Graphic Designer


NOTICE: The information in this book is true and complete to the best of our knowledge. All recommendations on parts and procedures are made without any guarantees on the part of the author or the publisher. Author and publisher disclaim any and all liability incurred in connection with the use of this information. We recognize that some words, parts names, model names, and designations mentioned in this book are the property of the trademark holder and are used for identification purposes only. This is not an official publication.

# Race Car Technology – Level Two

## Table Of Contents – RCT Level Two

| Introduction Section | Page |
|---|---|
| Lesson 1 - Goals of RCT - Level Two | 1 |
| Lesson 2 – Weight Placement & Goals | 5 |
| Lesson 3 – Geometry Settings and Considerations | 11 |
| Lesson 4 – Caster Settings | 21 |
| Lesson 5- Camber Settings | 25 |
| Lesson 6- Bump Steer | 31 |
| Lesson 7- Ackermann Settings | 39 |
| Lesson 8– Alignment – Front To Rear | 43 |
| Lesson 9 – Alignment – Rear w/ Roll Steer | 51 |
| Lesson 10 – Rear Steer Under Power | 57 |
| Lesson 11 – Toe – Front and Rear | 63 |
| Lesson 12 - Driveline Alignment | 67 |
| Lesson 13 – Spring Selection For Balance | 75 |
| Lesson 14 - Sway/Anti Roll Bars | 83 |
| Lesson 15 – Shock Selection – Matching Spring Rates | 91 |
| Lesson 16 – Bump Setups – Cause and Effect | 99 |
| Lesson 17 – Anti Design – Anti Dive and Anti Squat | 107 |
| Lesson 18 – Ride Heights – Documentations | 117 |
| Lesson 19 – Tire Selection and How To Read | 123 |
| Lesson 20 – Summation RCT Level Two | 137 |

# INTRODUCTION

Around 1992 I decided to change careers and become a race car engineer. My early experience in racing as a kid was spending countless hours at Daytona International Speedway in the pits, in the stands and around the mechanics, drivers and owners listening and hearing about how the cars handled. I was always fascinated by the design and setup of race cars, be they stock cars, road racing cars, formula or weekend SCCA cars.

I attended many years of races at New Smyrna Speedway and Barberville, now known as Volusia Speedway Park. I was interested in the "race", but always fascinated by the way the cars handled and how all of that was accomplished. I watched Dick Trickle prepare his car in 1975 at a friend's garage in preparation for the Speed Weeks show at New Smyrna and thought, I would like to be able to do that.

I am an engineer by education, degree and by nature and I knew someday I would have to get involved with racing. When that day came, I threw myself into the task of learning and inventing with more energy and determination than at any point in my life with anything I had ever done. It was a passion combined with a purpose. I was determined to find the truth about chassis dynamics and race car setup.

My work, and indeed my racing business, was born out of frustration and failure in trying to find really helpful information that I could use to set up a racecar. So, I set out on a journey that followed in no one's footsteps. Instead, I used one of my greatest personal assets, a profound and acutely developed ability to apply a common sense approach to problem solving. That is exactly what you will find in this book, a common sense approach to chassis setup, vehicle dynamics and race car design, together with solid engineering theory.

This is not a controversial race car setup book and agrees in principle with technology and theory taught in major college motorsports programs. To many, according to what I hear, the books of mine that have preceded this one have become their bible of racing knowledge. Much of the technology presented here has been more recently developed over just the last five years or so. For those who believe that we had already pushed the envelope of vehicle dynamics about as far as it could go by the early 2,000's, this book does not follow that line of thinking.

Regardless of what is on the pages, the proof is on the racetrack, and the methods in this book have been tested and proven to improve performance in race cars. They increase speed, improve basic stability and have already been used to win many races and championships in many classes of racing.

How then did these RCT books come about? When I began my career working with race cars, I found plenty of information on chassis theory, but I couldn't find conclusive information that would tell me how to set up my race car in the shop the right way the first time. I had to read between the lines and keep trying different setups, working by trial and error. I personally don't like trial and error. I want to be able to know exactly how to set up my race car and know not only how something works, but why.

This book will help you avoid the trial-and-error approach to chassis setup. It will teach you sound, proven technology that is both easy to understand and easy to use, so you can set up your race car in the shop and see the positive results on the track immediately, with very little tweaking. What follows is a common-sense approach to chassis setup, vehicle dynamics and race-car design, founded on solid engineering theory. However, you will need to have an open mind, and be willing to accept new ideas that may go against previous chassis setup thinking.

Just to make it clear, the technology presented here applies to all race cars, from quarter midgets to Formula One and everything in between. This book tends to lean towards stock car racing because it represents most of the world's automobile racing. But know that not only will it be useful for all forms of circle track racing from asphalt types to dirt cars, a great deal of the technology applies to all race cars.

Success comes at all levels of endeavor, and we can't all be champions. But we can all get better at what we do. The goal of this book is to give good, solid information that has been tested and evaluated and found to be the truth. It is not, and will never be, complete as long as we continue to push the envelope in the search for better performance, but it will lay the foundation upon which future race engineers can build their programs.

# Racecar Technology – Level Two
## Lesson One – The Goals Of RCT Level Two

Now that we know all of the parts and pieces of a race car, knowledge we learned in RCT Level One, we can get on with actually setting up a race car. In RCT Level Two, we will, in the logical order of importance, be setting up a race car. The methods we will use and the order we will progress in is mostly common with all race cars from short track street stocks to Late Models, to Prototype road racing, and even with the Formula One race cars.

Race cars don't just come ready to race. You might be buying a brand-new race car, starting to work with a used race car you just acquired, or maybe you are building your own race car. In any event, you must go through the car and setup each system of that car properly and in a specific order.

No one gets away with not doing any of this. You cannot be successful in racing if you don't work with these setup routines. And even if the race car manufacturer or previous owner says it is good to go, you cannot trust anyone but yourselves. We must verify each and every setup parameter of the race car.

If you are racing now, or have watched others race and have asked how they do it to make their cars so successful, it isn't magic and no one every stumbles onto a winning setup by accident. No matter how easy it all looks, a lot of effort and smart thinking goes into every winning race car. Here is what you can expect from RCT Level Two.

**What We Will Be Setting Up** – In this Level Two Course, we will go from simple, but necessary functions like determining ride heights. We'll set the weight distribution, and establish proper caster and camber for your application. We'll go through all of the various alignment settings and carefully explain why we are concerned with each phase and how those will make the car perform.

We will establish what spring rates to use, which sway bar will match those spring rates. We'll match our shock rates to the spring and sway bar rates too. We will choose tires and tire sizes and learn how to select tires and prepare them for competition. Then we'll put this package to the test and fine tune it at the race track.

When you get done with this Level Two Course, you will be able to take a modern day race car and set it up. When I first started into racing, I could find no good information to tell me how to setup a race car. Even today, information comes at us in parts and pieces and it is very difficult to put it all together into a winning combination. Now you can. This school is designed to provide all of the necessary information to guide you through the setup of your race car.

Our ultimate goals are as explained in RCT Level One. For those of you who forgot, or those who chose to begin your schooling with RCT Level Two, we urge you to go back and take the Level One course. In our introduction to Level One, we explained what our overall goals are for designing a high performance, winning race car. It bears repeating, so we will again explain.

The race engineer's goals are very simple in concept, but more complicated to carry out. On any race course, be it with circle track or road racing, on a dirt or asphalt surface, there are key areas of performance where your car needs to be made better than any of your competition and those are the following all associated with increasing grip.

*The goals for designing our race cars are basically the same be it a formula road racing car or a late model circle track car. Performance gains in the slower portions of the track cause all of the speeds to increase around the track including on the straights.*

**Maximizing Grip -** When non-chassis elements like powerplants are more or less equal, Grip is what will help you win races and championships. Motors accelerate you, brakes slow you, but Grip makes you faster through the slower turn portions of the track and relatively small gains in lateral Grip will produce huge gains in speed and performance.

What are the Key elements of Grip? The tire contact patch is where Grip is produced. Causing the tire contact patch to be larger increases Grip and causing the most vertical loading on that tire contact patch are the two key ingredients for maximizing Grip.

Maximizing the amount of traction that is available from the four tires on a race car, any race car, will make you as fast as you can be, all other things being equal. Everything we present in RCT L1 will ultimately lead to optimization of the race cars Grip and the use of that Grip to go faster.

## What are the basic parts that make up Grip?

- Loading On The Tire – The more load we can put on a tire, the more Grip that tire will have, period. But, this is gain in grip is not linear as we will explain later on. Load can come from the weight of the car, mechanical downforce from banking, and aerodynamic downforce, all of which will be explained in detail later on.

- Contact Patch Area – The greater the size of the contact patch, the more Grip. If we can find ways to make the tire contact patch larger, then that tire will produce more Grip.

- Tire Compound – The physical and chemical makeup of the tire can provide more Grip. The softer the material, the more Grip within limits. We have rules that govern the softness of the tires, but we need to stay very close to those limits.

- Load Distribution – A pair of tires on the same end of the car, or same axle, will produce more Grip when they are more equally loaded. The most Grip from a pair of opposing tires will come when they are equally loaded. There is a variation to this concept for dirt racing that will be addressed later on.

- Angle of Attack – What is called Angle of Attack, or Slip Angle, is when a tire is pointed slightly to the inside of the arc it is following through a turn. If a tire were to follow the exact tangent line around a curve or arc of the turn, it would not generate any side force to counter the centrifugal force.

In consideration of the other items that make up Grip, it is fair to say that none of those would be useful if it weren't for the creation of Angle of Attack. No matter what amount of Loading or size of the Contact Patch area is, or how soft the tire compound is, or how equal the load distribution between opposing tires, the car would not stay on the course without the tires developing an Angle of Attack.

So, there you have it. Those five things represent the parts that help make the Grip we seek to make us faster so we can win races. As we go through each part of the race car in the other Lessons, we will explain how to optimize of those parts to enhance our Grip and make us faster through the slow speed turns. And we will understand how that will in-turn make our whole lap faster at every point around the course.

## Why Does More Grip Make Us Faster?

When a race car turns, a lateral force called Centrifugal Force tries to push the car to the outside of the turn. The tire contact patches resist this force. The speed we can drive through the turn is limited by the amount of Grip we have in our tires. The more Grip, the faster we can drive through the turns.

The one often overlooked benefit of achieving faster turn speeds is this: the faster you exit a slow speed turn, the higher the speed at which you will start accelerating down the straight part of the track in-between the turns. So, speed gained in the slow speed turns will be carried down the straights too. It's not just that we gain speed in the slow speed turns with more Grip, we gain everywhere around a race track.

To give you an example, on a typical or average length Formula One track, one tenth of a second is about 17 feet on the race track. Some teams are a full second slower per lap than the fastest teams. That is 170 feet per lap that the faster cars move ahead of the slower cars each lap. In a fifty lap race, that equals 1.6 miles, or 2.5 kilometers.

*We often think of the formula one cars as being the ultimate race cars and perfect in every way. If that were so, then there wouldn't be a two second gap between the fastest and slowest cars. There must be a difference in designs that causes that huge gap in speeds. We like to think it is lack of mechanical grip that makes not only the slow turns faster, but the entire lap.*

We already know that some teams get lapped in a F1 race, and the average length of those tracks is around 3 miles. So, those lapped cars are on average 2.0 seconds per lap slower than the winning team. And they all run with the same choice of tires. That is hard to fathom.

Is the gain realized by the winning F1 teams all Grip? No, as we have read, some engine packages are down on horsepower and that is a factor in the difference in lap times. But what about the teams who have the same engine package as the winners? Why are they so slow? It could be that they lack mechanical grip in the slower portions of the race track.

Maximizing mechanical grip is one of the most important areas of race car engineering, even over and above aerodynamic grip. The reason is this; To gain maximum aero grip, you must have high speeds and the most gains from mechanical grip happen in the slower turns where the speeds and the aero downforce is lower for all of the teams. So, the only thing we can point to as the reason for the gain in performance for the winning teams is Mechanical Grip.

**The Concept Of Mechanical Balance -** If your car has more overall Grip than other cars early in the race, that Grip might not stay superior and you might end up being a slower car through the turns later in the race. This is due to the Balance Factor. There are two definitions of balance. They are not the same and they cannot be confused. One is handling balance which is defined as a car that is neither pushing (understeering) or loose (oversteering). Pushing is when the front set of tires has less Grip than the rear set of tires. Loose is the opposite.

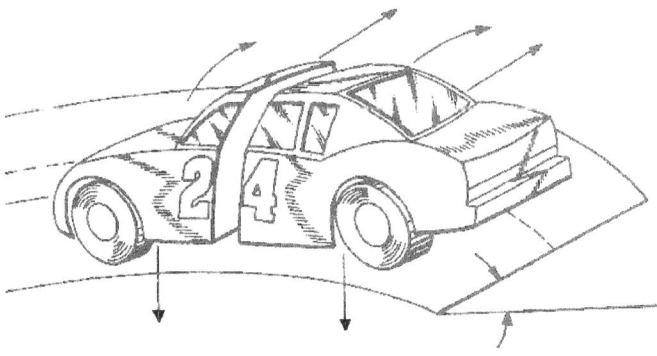

*The Mechanical Balance concept is simple. We want both ends of the car to work in sync so that one does not try to cause the other to do what it does not want to do. When the two ends of the car are working in unison, we can achieve the goal of mechanical balance.*

A team can attain handling balance fairly easily, but just the fact that the car is neutral in handling doesn't translate to speed or consistency. A car can have less overall Grip than other cars and still be neutral in handling. Although a neutral handling car is a goal, it is not the first and primary goal and not what we are discussing here.

The other balance is Mechanical Balance. Simply put, this is when the two suspension systems, front and rear, are working in sync and where the load transfer is predictable and maximized. This is what teams need to search for. The end result of all we do with chassis engineering must be towards the goal of Mechanical Balance.

*Circle track race cars are complicated, but much easier to setup than a road racing car. This is because the setup in a circle track car can be asymmetrical, meaning the spring rates, cambers, etc. from side to side can be different. The circle track car is only turning in one direction and so the setup need only be correct for that direction of turning.*

The Search and Desire for Mechanical Balance should be the foremost goal for every race engineer. A lot of good things come with achieving MB. The tires do not work as hard and will last longer. The car is much easier to drive. The car maintains higher turn speeds as the tires wear later in the race, or later in each stint for cars that are allowed to pit and take on new tires.

So, to reiterate what we have covered so far, our goals for this course are to gain knowledge, understand the terminology, and to learn how to setup a race car to maximize Grip in a way that utilizes a Mechanical Balanced state. In the Lessons to come, we will be presenting information you will need to understand and apply principles that will help achieve these two important goals.

And we are not leaving aerodynamic downforce out of the discussion, we just need to separate the mechanical Grip from Aerodynamic enhanced Grip. We need to do this because, many times in the search for the maximum aero properties, a team might deviate from the goal to achieve a Mechanical Balance. We think you can have the best of both worlds.

# Exam - In The Context Of This Lesson:

## The Primary Goal For Setup In A Race Car?

1) Make it easier to drive
2) Increase grip
3) Balance the two suspension systems
4) Create less chassis roll
5) 2 and 3

## Having More Grip Makes Us Faster?

1) On entry to the corner
2) On exit off the corner
3) Through the middle of the corner
4) Down the straights
5) All of the above

## Some Basic Elements Of Grip Are?

1) Contact patch optimization
2) Load Distribution
3) Tire angle of attack
4) All of the above

## Mechanical Balance Is?

1) Drive line angles
2) Tires are balanced
3) Suspension systems are working together
4) Weights are correct

## Lesson Two – Weights and Placement Goals

**Weight and Weight Placement Goals** – Part of the design of a race car is the establishment of corner weights, or how much of the weight of the whole car each tire ends up supporting. We use weight scales to measure the weight, or loading, under each tire. The amount of Static Corner Weight each tire supports statically is critical to how load ends up on the four tires through the turns.

Race teams measure weight with and without the driver in the car. If they just want to maintain certain percentages after making changes, they don't really need the driver in the car, they just need to know what the percentages are with and without the driver.

*Professional teams weigh their cars the same way we do. This Cup car at a test session at Daytona International Speedway is up on stands that are adjustable for height in order to make the scales level and on the same plane. The weights are critical to the handling of the car and are specific to the overall setup so that the load distribution on the four tires through the turns is ideal. These teams establish their weights and percentages with the driver in the car, and then record what they are without the driver. This way they don't have to chase down the driver in-between runs in order to get the baseline weights back after a spring or other change. The percentages taken without the driver are representative of those as if the driver were in the car.*

We need to establish and record these categories, or divisions, of weight: 1) Total Vehicle Weight, 2) Front and Rear Percentages of the Total, 3) Side Percentages of the Total, and 4) Diagonal Weight, also known as cross weight or wedge/bite in dirt racing.

If we are doing design work and computing for setups or calculating weight transfer, the entire weight of the car, including the driver, must be used because that is the weight the car will compete at. A race car will lose some of the total weight, change its cross weight and front to rear percent when the driver steps out of the car.

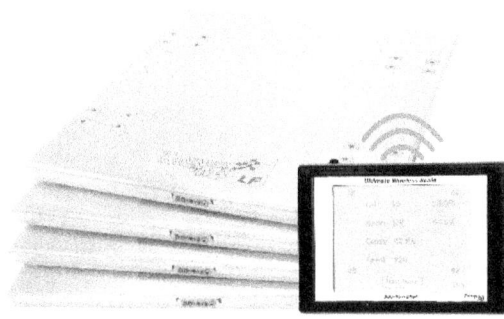

*Common wheel scales are used to find the weights, or loading, on the four tires. These individual weights are then calculated into percentages of: Front and Rear, Left and Right Side, and Cross Weight, or percent of the total vehicle weight that the RF and LR tires support. If we wanted to find Left Rear Weight, as is a common designation for dirt teams, we would subtract the RR weight from the LR weight.*

**Legal Weights –** Most sanctioning bodies and all tracks have weight requirements. The cars must weigh a minimum weight with, or without, the driver, have a legal left or right side percent maximum or minimum, depending on which way you are turning, and sometimes a minimum rear percent.

So, the first task in developing a weight plan for your car is to place any moveable weights in the car to meet the overall minimum total weight, and the side and front to rear percentages of that total. You might find that you need to move weight around in the car to meet those rules. Then later on you can adjust your corner weights to adjust for cross weight percent.

**Weight Distribution Related To Type Of Racing** - In road racing, the side percentages are usually the same, 50-50% as is the diagonal percentages being 50-50%. This is because a road racing car must turn both ways, left and right, and therefore must have a more symmetrical weight distribution.

In circle track racing, where the cars turn only one way, left in the U.S. and sometimes right in other parts of the world, we can create different side percentages and weights that are different from 50% diagonal, or what is termed cross weight.

In circle track racing, the use of the term Cross Weight gives us an indication of the load distribution on the four tires. It is defined as the combined load that is resting on the Right Front and the Left Rear scales added together and then divided by the total vehicle weight. That math gives us a percentage number to relate to, i.e. "51.4 percent of cross".

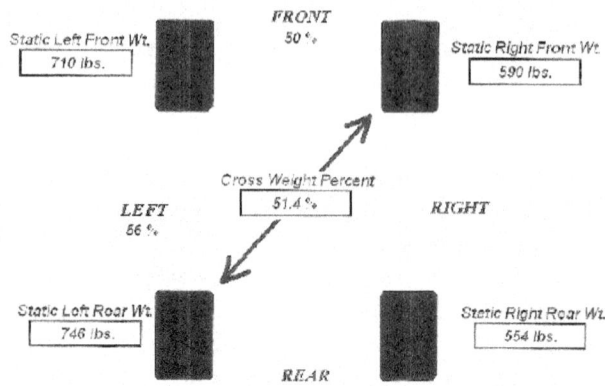

*Road racing cars must turn both ways, left and right. Therefore, the cross weight percent must be 50-50%, or the same for both diagonals. If it were different, then the load distribution for left turns would be different for right hand turns and the handling would be different.*

**Dirt Verses Asphalt Terminology** – For asphalt cars, the percentage number for cross weight is used, and for dirt cars, the terminology might be a little different. Most dirt teams refer to the amount of "bite" or "left rear weight" in the car, which is the number of pounds of weight that the LR tire supports over the RR tire. By subtracting the amount of weight that the RR tire supports from the LR tire weight, we arrive at a number and call that the amount of bite in the car, i.e. "100 pounds of bite or left rear".

Nonetheless, that bite number has a cross weight number associated with it. A low bite number relates to a cross weight around 50%, and a high LR weight or bite number puts our cross weight up around 56-58% most of the time. Coincidentally, low and high cross weight numbers in asphalt racing are in the same basic range.

*This is a typical layout of weights and percentages for an asphalt late model race car. The total weight of the car is 2600 pounds. The front to rear percent is 50% and the left side percent is 56%. This car running on a medium banked 1/3 to ½ mile track would require a cross weight of 51.4%. That is calculated as the weight distribution needed in order for the correct weight distribution to be placed on the four tires at mid-turn. If we can calculate the weight transfer during cornering, then back out what we need for static weight distribution in order to achieve those dynamic weights, then we would end up with this static cross weight for this car. The Left Rear weight for this car would be 746 minus 554 = 192 pounds of LR weight. Most dirt late models run a higher rear percent than 50%, so this does not represent a true LR weight for dirt cars, just an example of how to get that number.*

**How Cross Weight Changes** - It is important to note that if we increase the load supported by one tire, say the LR, we also increase the load supported by the diagonal corner of the car, or the RF tire. As we make that change, the opposite takes place with the other diagonal in that those tires (the LF and RR) experience a decrease in the amount of load they support.

So, for example, when we increase the Bite in the car (left rear), we also increase the loading on the RF tire, and at the same time decrease the loading on the LF and RR tires. You cannot change one tires loading without changing the loading on the other three tires.

**In The Past** – Historically, these numbers meant little to the circle track racer related to different setups, different types of race tracks, or overall weight distribution. All we knew was that a certain range of cross weight or bite worked for certain conditions.

Today we have an entirely new understanding of cross weight and the use of bite as those relate to the setups of the cars. Beyond knowing generally that more cross weight makes the car tighter at mid-turn and less tends to loosen the car, we can now know exact numbers to be used with a balanced setup.

For simplicity, we will refer to the weight distribution of the cars, whether dirt or asphalt, as cross weight. Dirt racers as well need to understand the relationship of changing the LR weight and how that affects the other three corners of the car.

*When we weigh the car at the track, the scales there may not be on the same plane or level. The readings for cross weight are often different than what we read at the shop on pads that have been corrected to be on the same plane. So, we can usually rely on the front to rear percent, the side percent and total, but not cross weight percent. So, just record what the track scales tell you for cross as a reference.*

Remember that the load distribution in the car changes as we encounter lateral forces associated with cornering. Some of the load carried by the inside (the turn) tires will transfer to the outside (outside of the turn) and the overall load distribution will change. We need to set the static weight distribution in anticipation of this load transfer so that the four tires will have optimum loading through the turns.

**Multiple Ranges** - There are different ranges of cross weight percent that will make your car neutral in handling. The same car at the same race track may be neutral in handling in the range of 48 to 52 percent and also be neutral in a range between 58 to 60 percent of cross weight. Dirt racers often change the LR from less than 100 pounds of bite to upwards of 300 pounds of bite. That represents a range of from 48 percent cross to 60 percent cross weight, the same as with the asphalt cars.

The actual cross weight, or overall weight distribution, number that will work for your car is dependent on the front to rear weight distribution, the type of track and the setup. Generally, the flatter asphalt race tracks with less grip work better using the high range of cross weight percent with the associated high left rear weight, while the same car will work better in the lower cross weight range at the higher banked asphalt tracks.

Many teams who run on asphalt tracks will try to run the high cross weight range that they have successfully developed for flat tracks when they go to race at the higher banked tracks. They would do much better as far as being more consistent by switching to the low cross weight range on the high banked asphalt tracks.

**Cross Is Related to Front to Rear Weight Distribution** - The exact amount of cross weight that will make your car neutral is directly related to the front to rear weight distribution. The greater the rear percent, the more cross weight a car must have to stay neutral in handling. Again, this is directly related to the anticipation of weight transfer so that at mid-turn, the weights on the four tires are correct.

### Cross Weight Related to Front Percent

| Front Percent | | | | |
|---|---|---|---|---|
| 46% | 48% | 50% | 52% | 54% |
| 59.1% X | 55.6% X | 52.1% X | 48.6% X | 45.0% X |
| 54% | 52% | 50% | 48% | 46% |
| | | Rear Percent | | |

*This chart is a general indication of how the ideal cross weight will change when we change the front to rear percent. These numbers are for a particular car and not necessarily right for every car. But they do provide an interesting view of the relationship. One very interesting thing to note is that with a race car with 50% cross weight, we would need to run a rear percent of 49.4%. If a road racing car ran anything but 49.5%, then the handling would be off if that car were dynamically balanced.*

For dirt racing, with the associated higher than 50% rear weight distribution, the range of cross weight that will make the car neutral is opposite of that associated with asphalt cars. The high range is more normal to the high rear percent cars, while the low range also can be also yield neutral handling.

*We adjust our weight distribution by adjusting the loading on the springs, or the spring height. This coil-over setup has a threaded body on the shock and a ring that can be adjusted for height to compress the spring to provide the correct loading. Changes in spring loading will also change the ride height of the corners of the car. Changes in wheel weights must be made at all four corners of the car at the same time. We'll explain how to do that.*

**Sway Bar Effect** - We can cause the static Cross Weight to change by pre-loading the sway bar. Pre-load is when we twist the bar at static ride height through an adjuster mechanism. Pre-load on the sway bar puts load onto the RF and LR tires which increases the amount of static cross weight percentage. The larger the sway bar, the greater the effect of pre-load on the car.

If we want to pre-load the bar, we should know how much the cross percentage will change, then lower the static cross weight percent by that amount. Then when we pre-load the bar, we will have the exact static cross weight we desired and want, not something higher. The pre-load on the sway bar mostly helps provide bite off the corners and shouldn't be used to tighten the car through the mid-turn phase.

**Weights Can Indicate Balance** – When you run a setup that is balanced dynamically, that means the front and rear are working together. This also means that the load transfer is predictable and your car will need a predetermined cross weight percent in order to end up with the correct weight distribution at mid-turn. The cross weight percent we need to run in our cars to make them neutral in handling can be an indication of how balanced our cars are.

If the ideal cross weight for a particular application that is dynamically balanced is 51.4 percent, and we find that we need to reduce cross weight to 49.5 percent in order to be neutral in handling, then our setup must not be truly balanced.

As we change our setup from unbalanced to a more balanced setup, we will also need to change the cross weight percent to what works best for that dynamic balance. If your team insists on using a fixed cross weight percent that you "always ran", then your setup cannot be improved until you decide to change the cross weight along with positive chassis setup changes. Weight distribution and balance go hand in hand.

**Weighing the Car In The Shop** – When we setup our scales at the shop, or wherever we do it, we must level the scales. If the floor is relatively level, say in a commercial building, things will be easier. If you are in a garage intended to house a car, the floor will usually slope towards the garage door. This allows any spilled water to flow out the door and not accumulate.

*In today's racing, it is necessary to have a set of scales and a good set of stands in order to maintain your wheel weights. Note that if you make changes in spring rates, camber settings, sway bar pre-load, etc. those changes will also change your race cars weight distribution.*

It is less important for the scales to be level front to rear, or side to side. It is extremely important for the scales at each axle to be on the same plane, just like we talked about in the Ride Height Lesson. So, if we have put the scales on the same plane and the front to rear were off level by less than an inch, it's not the end of the world. Side to side we can be off level by up to a half inch and be alright.

Weigh the car with all of the fluids and driver weights it will have in it when it is on the track. Many teams will simulate driver weight by putting lead in the driver's seat, but when doing this, you should put about 30% of the lead in the foot well and the other 70% in the seat. And you can place lead on top of the fuel tank to simulate a full tank. Fuel weighs about 6 pounds per gallon.

Bring all of the tires up to race pressures because they will grow differently front to rear from cold pressures and affect the height of the wheels which will affect the loading on the scales. And make sure the front wheels are pointed straight ahead.

If you have high rebound shocks on any corner of the car, wait till they stabilize before you take readings. For adjustable shocks, reduce the rebound all of the way while you are scaling the car. Disconnect the sway bar while you are taking scale readings. You can load and/or preload the bar later on. Now you can read the scales.

**How To Change Weight Distribution** – Changing the loads can be done one of two ways. Either you want to move or add weight, or you want to redistribute the weights that are already in the car to different wheel loadings. Moving weight or adding weight will change the loading on the four tires and change the front to rear percent, side to side percent and the cross weight percent.

If you want to change load distribution, but not side or front to rear percent or the total weight, there is a way to do that and not change the ride heights appreciably. I say that because there will always be a little tweaking needed to fine tune your ride heights after you have changed load distribution.

If you need to increase the cross weight (RF and LR increased as a percentage of total weight of the car), you will force down the right front and left rear springs and ease up on the left front and right rear springs. The most common device for doing that is a screw jack or adjuster rings on a coil over shock.

So, if the springs are similar in rate from side to side like at the front, you can add turns to the adjuster at the RF in the same amount as you take turns away from the LF spring. This same process can be used on a road racing car where the rear springs are the same. At the rear on a circle track car, the rear spring rates are usually different side to side. If you divide the softer spring rate by the stiffer spring rate, you get a percentage number.

If the LR were a 200ppi spring rate and the RR were a 400ppi rate, the percentage would be 0.50. If you want to turn the LR spring down by four turns, you would turn the stiffer RR spring the opposite direction by the percentage times that, or up two turns. That is because the same number of turns on the stiffer spring will move that corner twice as much as that number of turns on the softer spring.

Try to record how much the cross weight changes per turn of the adjuster as you make changes to the front and rear spring loads. That way, when you want to add cross at the track, you'll know exactly how much you need to turn your adjusters for the percentage of change desired.

After you have established your desired weight distribution and if one end or the other is higher or lower than the established ride height, just turn the adjusters for each spring on that end the same amount to get the ride heights back. That way the cross weight won't change.

If you follow this rule, your ride heights will remain close to what they were when you started. Now you can hook up that sway bar. If you want to preload the bar, do it now and see how much cross percent it adds per turn. Then you can make adjustments to return to the original cross weight if you add turns at the track.

**Bump Setups and Cross Weight** - If you are running the bump setups, when the car goes down onto the bumps, there is a good chance that your sway bar will gain or lose loading and affect the load distribution on the four tires. This would change your handling if not corrected.

Set the bar to neutral at ride height. Push down on the car till it is on the bumps and then note how many turns it takes to go to neutral again. If this gain or loss is not accounted for, then when we race the car with the front or rear down on the bumps, our weight distribution will be something other than what we want and need.

Also with bumps, if the bump device is not spaced correctly from the shock, when the two front shocks contact the bumps, the cross weight may change significantly. It is a good idea to remove the front springs and let the bumps support the front of the car. Then check the cross weight on the scales. If it is different than when on the ride springs, add or take away packers, or spacers, till the cross weight is the same as when you set it on the ride springs.

**Weighing At The Track** – The officials track scales are almost never level. Get your weight percentages at the shop and trust those numbers. When you get to the track, re-weigh the car on the track scales and note the difference in cross weight percentage and maintain that difference.

The front to rear and side to side percentages should not change much at all because we stated at the top of the article that it would take more than an inch or so of difference in level side to side, or slope in the floor, to make a noticeable difference in readings. It is with the cross weight percent readings that the unlevel scales really show a difference from your shop scales.

**Summary** – Before getting into the setup of the car, we need to establish our race car weights being: 1) Total Weight to meet the rules, 2) Front to rear and side to side Weight percentages, 3) Weight distribution on the four tires, or cross/bite numbers.

We need to maintain those numbers when we change cambers, pre-load the sway bar, make spring changes and/or scale the car at the track. Consistency starts with maintaining the weight distribution in the race car.

# Exam - In The Context Of This Lesson:

### Weight Distribution Includes Which?

1) The weight on the four tires

2) The front to rear percent

3) The side to side percent

4) The diagonal weight percent

5) All of the above

### Before I Set Cross Weight, I Need To?

1) Set the total weight of the car

2) Set the front to rear percent

3) Set the side to side percent

4) All of the above

### We Set Cross Weight So That?

1) The cross equals the left side percent

2) The cross weight is the same as the rear percent

3) The cross weight is 50-50 for road racing

4) The cross weight percent causes ideal dynamic tire loading

5) 3 and 4

### "Bite" is The Same As "Cross Weight"?

1) True

2) False

### The Following Can Change The Weight Distribution?

1) Sway bar pre-load

2) Front end travel

3) Travel onto the bump stops or springs

4) Camber Changes

5) All of the above

## Lesson Three – Geometry Settings

In this Lesson, we will talk about double A-arm geometry, and on circle track and some road racing cars with rear solid axle suspensions, this will be only on the front end. For formula road racing cars, it will be front and rear. Geometry can cover many different aspects of the construction of the chassis, but for the purpose of this Lesson, we'll only be involved with what is historically called the roll center, or what I always like to call the Moment Center.

The reason I call this point the moment center, as stated in RCT Level One, is because it represents the bottom of the moment arm for the AA-arm suspension, being the front in a circle track or road racing car, and possibly the front and rear in a formula car.

*The moment center, or what is typically called the roll center, is a point formed by the angles of the upper and lower control arms. The location of the Moment Center moves as the chassis dives and rolls through the turns because those actions cause changes to the control arm angles. We take measurements from the ground and from the centerline to find the location of the MC.*

**Theories About AA-arm Geometry** - After much research and testing over the recent past by myself and others, it has been determined that the primary use and function of moment center design is to optimize the camber change and contact patch size in a AA-arm suspension, as well as to influence the roll resistance.

Other theories exist that talk about Jacking Force and that has also been tested. The conclusion the testing reached is that the control arm angles that produce the best camber change and contact patch area for both the inside and outside (the turn) wheels are basically the same for moment center and jacking force theories.

In RCT Level One, we told how the moment center is arrived at and what makes up the parts and pieces of roll/moment center geometry. We won't re-speak that Lesson here and if you for some reason did not enroll in the RCT Level One course and don't know how the moment center is arrived at, then I urge you to go back and take that Course.

What we will be learning here in this Lesson is how to design a better moment center and what works for different types of race cars. I can tell you through many years of research, consulting and testing that what you will learn here is not widely known in the industry, even at the highest levels of competitive racing. But it is very effective in enhancing performance.

**Two Moment Centers** – There are two moment center locations we are concerned with. First is the Static MC location and that is where the MC is located when the car is at ride height and still. The other is the Dynamic MC location, being where it moves to as the chassis moves to a new attitude going through the turns. The dynamic position is the most important location to design for because it influences the camber change and the contact patch area.

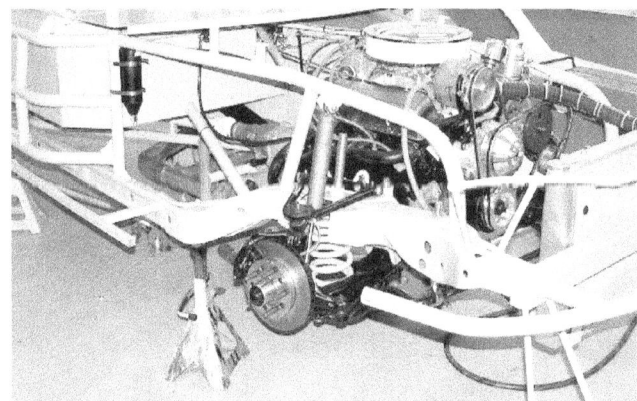

*Stock type of cars can benefit from moment center re-design too. They are especially prone to poor design due to being stock and intended for street use. You can cut off the upper arm mounts and create a plate type of mount that is easy to adjust for camber and moment center. This example stock clip front end has an aftermarket upper control arm mount welded to the stock frame. If allowed in your rules, this makes mounting an aftermarket upper control arm and adjusting your Moment Center location much easier.*

We can look at both the static and dynamic MC locations in readily available geometry software programs. To find the MC location, we need to know the height and width of each point that makes up the AA-arm geometry for the suspension we are working on. Those points include the two ball joints on each side as well as the top and bottom chassis mounts for the control arms.

Those points create the control arm angles that in-turn form the MC location in the system. The MC will move once the chassis starts to move in dive and roll because those control arm angles will change along with that movement. So, we can track where the MC moves to with the chassis motion.

**Circle Track Car Design** – For circle track cars, we want the MC to start out to the inside of the turns and for cars turning left, that would be left of centerline from a driver's view. We also do not want the MC to move very far in most cases, although some experimentation here could help some situations.

If we get this design right, the camber change in the outside wheel (right front for left turning cars) will be very little. Race tires do not like camber change, especially at the right front, or outside tire. If the camber of that tire stays relatively unchanged through the dive and roll motions of the chassis, then that tire will have the most consistent level of grip possible.

As to the inside wheel/tire, there will always be camber change and the best we can do is minimize the amount of change. Designing the MC to be left of centerline in a left turning car does produce control arm angles that result in minimal camber change for both the inside and outside wheels.

**Road Racing Car Design** – In road racing, the cars turn right and left. It is also true that there is a combination of control arm angles that will produce the least camber change resulting from the cars chassis movement in dive and roll going through the turns.

If we get this right, there will be minimal camber change for the outside tire, but still like the circle track cars, the inside tire will always experience camber change. The best we can hope for is minimal camber change in the outside tire to maximize that tires grip and then we have to live with whatever grip results from the inside tires resulting camber.

**How To Design The MC Location** - If you want to know exactly where your moment center is located in both the static (at ride height and not moving) and dynamic (at speed going through mid-turn) locations, you will need to purchase a front end geometry software program. It's just not feasible to draw it out on paper. This discussion will assume you are using, or plan on using a program. If you don't ever get to the point, at least here you will learn how that is done and that information is important to any racer.

*In order to find your Moment Center in both the static and dynamic locations, you will need a computer program. Teams have tried in the past to draw the MC out on a large sheet of paper, but drawing the dynamic location is very difficult, if not impossible to do. If you don't have a program, you might find a friendly team who does and use that to find your MC location.*

Because I have been doing this for over twenty years now, I know that there is no trial and error method that works to find the best geometry design. The system is just too complicated. If you don't have one of these programs, or cannot afford one, see if a fellow team will work with you and let you use theirs. The idea of that may be a stretch in some racing circles, but most racers want to beat you on your best day, or not.

**Getting Started With Measurements** – First you will need to get measurements. We measure the height and width of each chassis mounting point and the centers of rotation of the ball joints on each side. Once you have your numbers, you will enter them into the program. The two important sets of numbers are height and width of the (1) static MC location, or SMC, and (2) the dynamic location of the MC, or DMC.

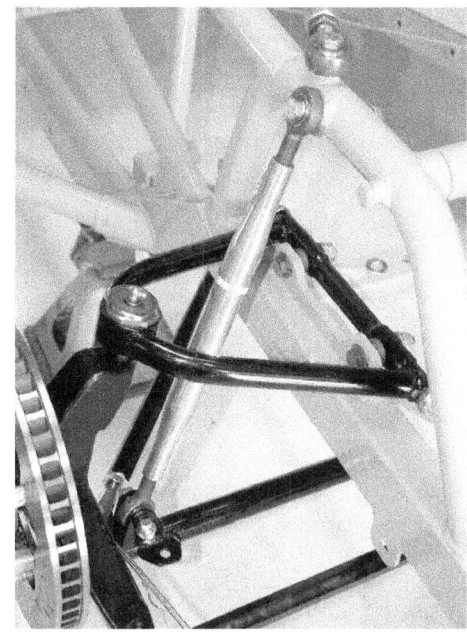

*When measuring your car, make sure to fix the spindles in the same location as they would be at normal ride height with all of the weight in the car. This well made link that replaces the shock in a late model is adjustable for length and makes positioning the spindle quick and easy once the wheels are off and the car is raised up.*

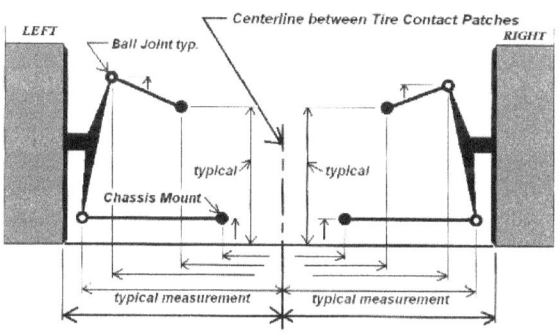

*To find the MC location, we measure the height and width of each point used to establish the intersections. These are: 1) the ball joints on each side including the upper and lower ball joints, 2) the chassis mounting points for the upper and lower ball joints. The centerline we will use is the midpoint between the two tire contact patches.*

The location of the DMC is most important because this is where our front end is doing all of the work. Over years of practice, experimentation and use, we have narrowed down the range of location of the DMC that most racing technology experts will agree works the best.

Coincidentally, what moment center advocates think is a good location for the DMC agrees with the proponents of the jacking force theory. Since we are all in agreement on the location, and not necessarily why that place is good, we can continue without controversy.

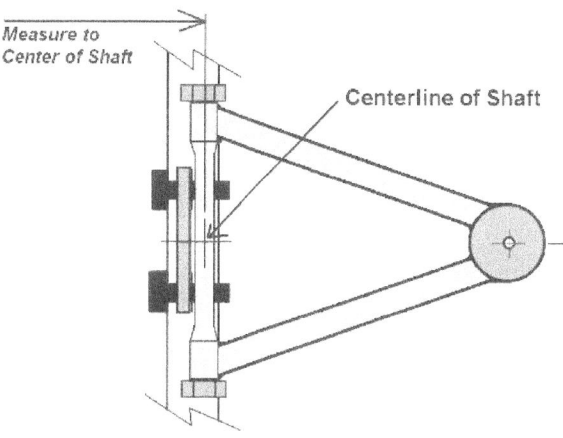

*When measuring the upper chassis mount on a typical circle track stock car or some road racing cars based on those cars, we measure to the centers of the bushings, or centers of rotation. Some shafts are offset between the bushings, so be careful when you measure to not mistake the center of shaft at the middle for the center of rotation.*

Not only is the location important, but the distance the MC moves from SMC to DMC is important. I always say the MC moves fairly quickly from SMC to DMC, but I still want the two locations fairly close together. This means that I want them less than eight inches apart. Yes, cars will work if they are farther apart, but may not work as well, especially on entry.

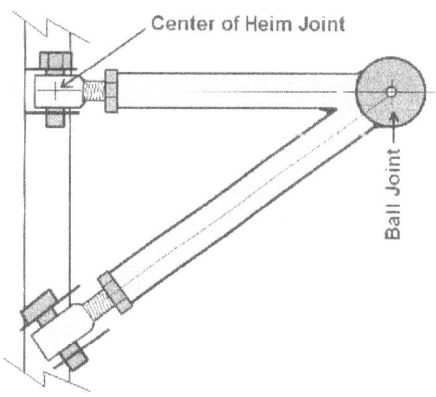

*For upper control arms that are this shape, you can just measure to the chassis mount that is ninety degrees to the centerline from the ball joint. With the limited angular movement of the wheel for most applications, this is adequate for finding the MC location. For stock lower control arms where the line between the lower mounts is angled from a top view, the front mount is usually near ninety degrees off of centerline inline with the ball joint and you can also use just that mount for MC location.*

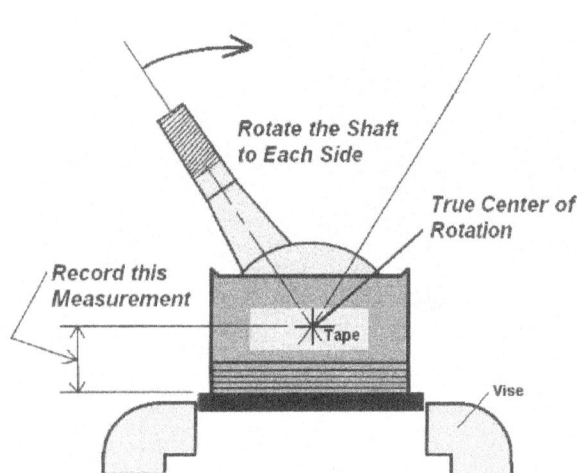

*To find the center of rotation for a ball joint, you can go through this process. Mount the ball joint in a vise or similar holder, and rotate it all of the way to one side. Use a straight edge to project the center of the shaft onto a piece of tape on the side of the housing. Rotate the ball joint all of the way to the opposite side and again draw a line through the center of the shaft onto the housing. Where the two lines cross is the center of rotation and you can measure from the edge of the housing to that mark to use for all similar ball joints.*

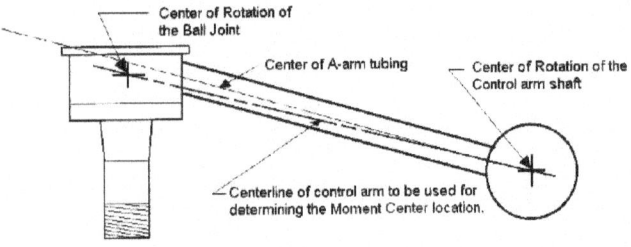

*The upper, or lower for that matter, control arm angle is formed from a line projected through the centers of rotation of the ball joint and chassis mounting point. This line almost never follows or represents the center of the connecting shaft that joins the ball joint to the chassis mount.*

**Lower Control Arms–** The lower control arms determine somewhat the lateral location of the MC, but mostly they determine the amount the MC moves from SMC to DMC. So, the angles are important. So we usually set these angles and forget them. Most of our work in MC design is done with the upper control arm angles.

*To make changes to the lower arm angles, most modern race car designs utilize slotted mounting holes. When you find the best angles for your use, tack weld the washer in place so that it doesn't move during use. You can also use different length ball joint shafts, like we used for the upper arm angles, to change and perfect your lower control arm angles.*

From experimentation, for circle track cars, we have learned the following: A) Both lower control arms need to be positive, that is, angle with the chassis mount heights lower than the ball joint heights; B) you need to have more angle in the left lower control arm than the right one; C) The angles need to be between 1.0 to 2.0 degrees different, no more, no less.

What the proper design for MC does is limit the movement of the MC's to under eight inches from SMC to DMC in most cases. The reason we don't want the lower control arm angles to be negative (ball joint lower than the chassis mount) is mostly because that affects our camber change in a negative way

**Upper Control Arms –** The angle of the control arm is the angle from horizontal of a line formed by the centers of rotation of the chassis mounts and the ball joints, not the angle of the tubing connecting the two. The angles of the control arms dictate the locations of the SMC and the DMC. The upper control arms mostly determine where the MC's are laterally. We change these angles to move the lateral location of the two MC's. Here is how.

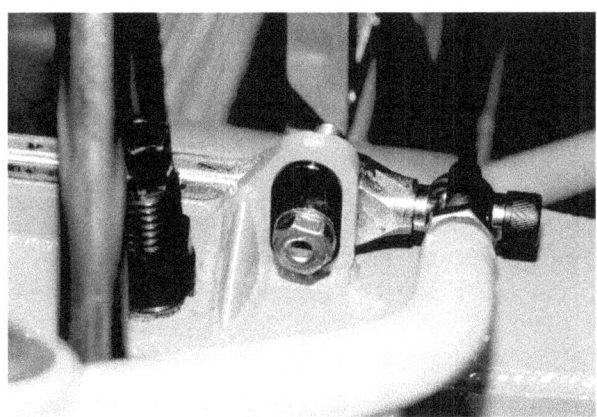

*Most upper control arm mounts are adjustable for height. This one has slugs you can use in different offsets for the hole to move the chassis mounts up or down. You can also change the ball joint shafts to raise or lower the ball joint center of rotation. Either will cause changes to the upper arm angles and move the Moment Center to where you need it to be.*

The MC location is determined by the location of the instant centers, or IC's. We don't necessarily need to know where the IC's are located, we just need to know that when we change the angles of the uppers, we are changing the locations of the IC's. This again is all covered in RCT Level One.

Generally, putting more angle (degrees measured from horizontal) in either upper control arm will move the MC towards that side. So, if we increase the left upper control arm angle and decrease the right upper control arm angle, the MC will move to the left. That is because we have moved the IC's on both sides in that direction.

*This design has only two mounting holes that don't provide enough fine adjustment to design your MC location. To change the arm angle, you can either change the height of the chassis mount to the other hole or change the height of the ball joint. You can install ball joints with different length shafts to raise or lower the ball joint center of rotation. Some modified cars use mono-ball ball joints that are easy to make height changes with. That is the essence of MC re-design associated with the upper control arm angles.*

*Modern aftermarket, or non-OEM ball joints can be ordered with different length shafts. This way you can move the center of rotation up or down to fine tune the arm angle and ultimately the MC location. Just the shaft can be purchased if you already have these type of ball joints and don't need the entire assembly.*

*Many dirt late models and some road racing race cars use square tubing uprights to mount the upper control arm to. On these designs, to change the arm angle, we only need to drill new holes into the uprights at the location that will produce the desired arm angle we need.*

**Road Racing MC Design** - For road racing cars, the upper and lower control arm angles will be the same left to right. There are exceptions to this rule, but for the most part, they are the same. For optimum design, a lower MC works best and a MC that moves very little from centerline through dive and roll works best. That combination produces the best outside tire camber change, or what we call minimal camber change.

*Road racing cars can benefit from MC redesign too. The angles of the upper and lower control arms will determine the camber change that takes place during cornering. With wide tires like those used on this class of prototype car, there is little room for error with the tire cambers.*

The resulting upper control arm angles will be between 10 and 20 degrees in most cases for typical designs. The exception is in the Formula One cars where the arm angles are reversed with the ball joint lower than the chassis mounts. This is a special exception that we won't go into in this Lesson because those cars represent a fraction of the race cars in the world.

**Camber Change Considerations** – During early research into the front end geometry and the MC locations, we were working with conventional setups. The soft springs combined with bump technology we see today was still some ten years away. What we learned was that there was a combination of upper control arm angles that would yield near zero change in the outside tire camber through turn entry, middle and exit off the corners.

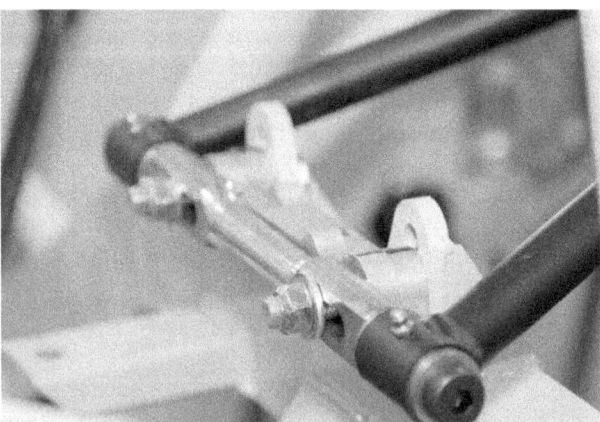

*You can adjust your camber easily on most race car designs. This upper mount on an asphalt late model has spacers between the upper control arm shaft and the upper chassis mounting plate. When you optimize your MC design to reduce the camber change, don't forget to adjust the static camber. See the Camber Lesson to learn more about the correct way to do that.*

With the modern circle track setups, the front ends travel more vertically now than before, so zero change is not now possible. The right front always gains negative camber. The left front wheel always lost much of its positive camber with the old school setups and still does with the newer soft spring setups. But now with a lot of front end travel, it loses even more camber and the right front gains a lot more negative camber. We cannot do much about that, it's the world we live in now days.

The positive thing about the bump setups is that the cambers, once the chassis has moved down onto the bumps, stay relatively the same with little change. Remember what we said about the older conventional setups, they did not like camber change. We now have systems that again produce little camber change.

**Ideal Moment Center Locations** – Now that we know how to re-design the MC and a little about arm lengths and angles, let's discuss where we need to end up with our DMC. We always talk about the DMC location because it is the only one that counts. Mid-turn dynamics represents 80-90 percent of our focus with chassis setup.

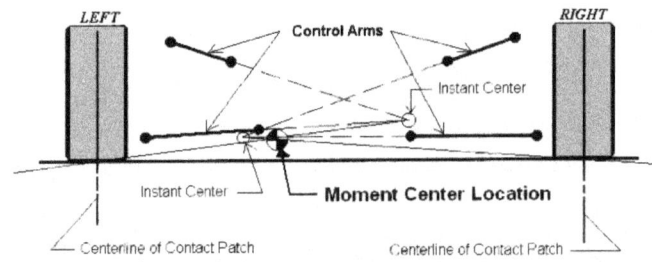

*The MC is positioned at the intersection of the lines from the Instant Centers to the center of the contact patches. You can see in this illustration how that is done. The control arms change angles when the chassis moves in dive and roll. So, the lines change as does the MC location. For most applications for circle track racing, the MC is best located towards the inside of the turn. For road racing, we want the MC to stay very close to the centerline of the chassis through dive and roll.*

The ideal MC location varies by class and type of racing. For street stock and most of the more conventional setups, the DMC needs to end up somewhere under the center of gravity. So, if you have a 60 inch track width and are running 54 percent left side weight, the DMC should be located left of centerline by about 2.4 inches.

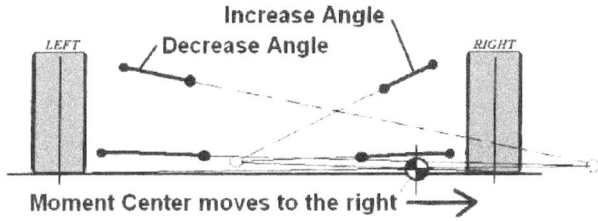

*We can move the location of the MC by making changes to the control arm angles. Putting more angle in an upper arm, in this case the Right Upper arm, will move the MC location to the right. Decreasing upper control arm angle in an upper control arm, in this case the Left Upper arm, will move the MC towards the opposite side, or right side. This give you some idea of what to change to move the MC in the direction you need it to move.*

You can experiment with the location by moving it farther left, but on those cars, with the high center of gravity, you can definitely overdue it. Keep the MC left of centerline, but not too far left and your car will turn very well, especially the dirt cars that are stock based.

For late models on dirt or asphalt, we see a much more aggressive approach to MC location. The DMC for soft conventional setups follows the above example in reference to the center of gravity. So, for a car with 66 inches of track width running 58 percent left side weight, the DMC should be located about 5.5 inches left of centerline to put it under the CG. This used to work very well and still does for those cars.

As to the super late models running on bump setups, the design can place the MC a little farther left for several reasons. One, with the high amount of dive and stiffness in the spring rates of the bumps, we can venture much farther left.

For most bump late models, the common location for the DMC is from 12 to 16 inches left of centerline. Most modern and successful teams, if they were honest about it, will tell you that this DMC location is where the car turns best and coincidentally that is where we find the camber change is the least.

For those types of cars, the fact that the car is down on the bumps when entering and going through the turns, there is very little vertical movement (chassis dive) and very little chassis roll, so the camber change is minimal regardless of the MC location. Nonetheless, we still design for the locations we stated above.

For road racing cars and formula cars, the best MC design has the MC starting at centerline in static location and then moving very little from centerline through chassis dive and roll. And, with minimal camber change in the outside tire. In early experimentation with both a AA-arm front and solid axle rear type of cars and later on with prototype cars designed with AA-arm front and rear suspensions, optimization of the control arm angles resulted in significant gains in grip through the turns.

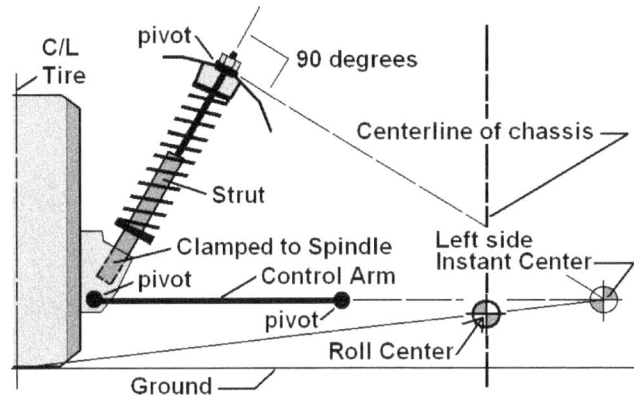

*Stock type McPherson strut designs have Moment Centers too. The location is a little different than with a AA-arm suspension. The lower arm forms a line to the instant center, and a line at ninety degrees off the line of the shock, or strut, forms the other line to the instant center. These systems are almost always symmetrical, and so the static MC is always at centerline, or very close to it. If the cambers of each wheel are adjusted by moving the upper mount of the strut, then the line off the strut we use to find the instant center also moves causing the MC location to move. For Circle Track racing, we need more positive camber in the inside wheel and more negative camber in the outside wheel. This necessarily moves the instant centers and the MC location to the outside of the turn, which is not ideal by any means. This is why these cars are very hard to make work.*

**Summary** – The geometry design is one of the very first and most important design features on any race car. Other geometry changes such as camber and caster change won't appreciably affect the MC location, so we can do this design early on in our overall chassis setup routine.

Changes made to move the MC location will affect the bump steer and other design parameters because those may be affected by the control arm angles. So be careful when you make changes to the MC that you don't inadvertently change other geometry settings.

I can tell you without a shadow of a doubt that MC design and optimization has the greatest effect on chassis related performance gains aside from setup balance and load distribution. Remember that every race car needs proper loading in the tires and maximum tire contact patch along with that loading. And your MC design helps achieve the best contact patch design.

# Exam - In The Context Of This Lesson:

### The Moment Center Is Called That Because?
1) It only takes a moment to find it
2) It is the bottom of the moment arm
3) It is different than the roll center
4) It is the center of the chassis roll

### The Two Primary Moment Center Locations Are?
1) Right of Centerline
2) Static location
3) Dynamic location
4) Left of centerline
5) 2 and 3

### For Circle Track Racing, The MC Should Always Be?
1) Towards the outside of the turns
2) Towards the inside of the turns
3) Higher than the ground
4) Lower than the ground

### For Road Racing, The MC Should Always Be?
1) Towards the outside of the turns
2) Towards the inside of the turns
3) Above Ground
4) Close to centerline

### What Are The Primary Measurements Needed To Locate The MC?
1) Distance from the centerline
2) Height off the ground
3) Distance from the front axle
4) 1 and 2

### The Lower Control Arms Mostly Influence?
1) Where the MC is located laterally
2) The height of the MC
3) How far the MC moves static to dynamic
4) The camber change of the wheels

### The Upper Control Arms Mostly Influence?
1) Where the MC is located laterally
2) The height of the MC
3) How far the MC moves static to dynamic
4) The camber change of the wheels
5) 1 and 4

### We Can Change The MC Location By?
1) Raising or lowering the chassis mounts
2) Using taller or shorter ball joints
3) Changing the lower control arm angles
4) Changing the upper control arm angles
5) All of the above

### Putting More Angle In The Left Upper Control Arm Moves The MC?
1) Towards the right
2) Lower to the ground
3) Towards the left
4) Higher off the ground

**Putting Less Angle In The Left Upper Control Arm Moves The MC?**

1) Towards the right

2) Lower to the ground

3) Towards the left

4) Higher off the ground

**We Can Find The Control Arm Angles By Measuring The Angle Of The Tubing?**

1) True

2) False

**The Most Positive Thing About The Bump Setups Is?**

1) We don't need shocks anymore

2) The car is lower

3) The cambers don't change very much

4) The car doesn't roll

## Lesson Four – Setting Caster

Caster settings are set relative to the type of car and the design of the race track as well as the drivers preference. When you determine the correct settings, you need to maintain those settings. We will discuss what affects caster has in our race cars as well as how to properly measure it.

It is important to know how to properly measure for the amount of caster in your race car. All teams need to learn the proper procedure for determining the correct caster that exists in the front end geometry of their race cars because it is many times a driver preference. Here is the definition of caster.

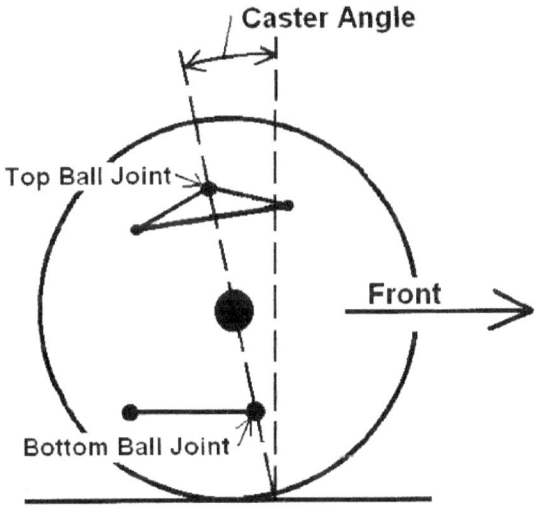

*One of the most basic, but also most important, settings on the front of the car is Caster. This setting affects the drivers feel when steering the car through the turns. Caster occurs when the upper ball joint is mounted to the rear of the lower ball joint. This creates the caster angle. Positive aster in the front wheel assembly is created when the ball joints are offset, from a side view, so that the upper ball joint is farther to the rear of the car than the bottom ball joint. The degree of Caster is related to the angle in degrees that a line through the ball joints forms from a vertical line.*

**Caster Defined** – Caster is a design condition that serves to cause a wheel to want to track straight ahead. A common example is a bicycle front wheel and fork assembly. The tube that the handlebars are attached to is mounted in a set of bearings above the fork and from a side view this tube is angled so that the bottom bearing is ahead of the top bearing. If we turn the front wheel to the side from straight ahead, it will want to return to straight ahead by the effect of caster. This is called positive caster. The same effect is present in the front wheel assemblies of our race car.

The car wants to return to straight ahead because when the wheel is turned, the positive caster effect raises both the bike and your race car. The weight of the car pushes down on the wheel and it naturally wants to return to straight ahead where the car is at its lowest.

**What Caster Split Does** - To ease the amount of effort it takes to turn the wheel in our circle track race cars, we introduce caster split into the design. Split means that we set different caster amounts into each front wheel assembly so that the car will want to turn only one way naturally and thereby reduce the amount of effort it takes for the driver to hold the steering wheel when negotiating the turns.

Proper split for circle track racing cars turning left means that the left front wheel will have less positive caster, or in some cases negative caster, as opposed to the right front wheel. Teams have been known to set negative caster in the LF wheel and positive caster in the RF wheel. If the driver took his/her hands off the steering wheel, the car would naturally turn left due to the caster split we have described.

For road racing cars, the caster amount will be equal on each side of the front suspension so the steering effort will be the same for right and left hand turns. The trend in road racing is to run greater amounts of caster because of the camber change that caster produces, gaining negative camber in the outside wheel and gaining positive camber on the inside wheel. These effects produce better contact patch shapes for greater front grip.

Because road racing cars have equal caster side to side, all discussion about caster split will relate to circle track cars that turn in only one direction.

**How To Measure Caster** - To measure caster in each wheel, we use a Caster gauge. This tool attaches to the wheel hub. To check the amount of caster, we need to follow these instructions:

1) Attach the Caster gauge to the Right Front wheel hub first.

2) Turn the steering wheel to the right so that the right front wheel has turned exactly twenty degrees from straight ahead.

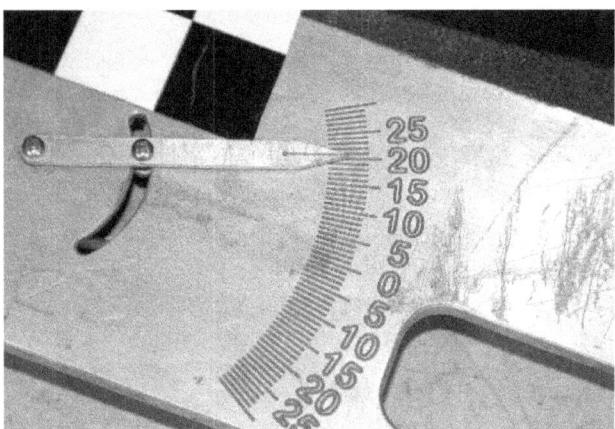

*To measure caster with a bubble gauge, level the gauge and set the bubble at zero on the Caster side of the gauge with the wheel turned right exactly 20 degrees. Some teams use one edge of the bubble so that reading the angle is more accurate than trying to estimate the center of the bubble.*

*It is a good idea to use turn plates in order to know exactly when you have turned the wheel twenty degrees, the standard amount for caster measurements according to SAE standards, to which the gauges are designed to accurately measure caster.*

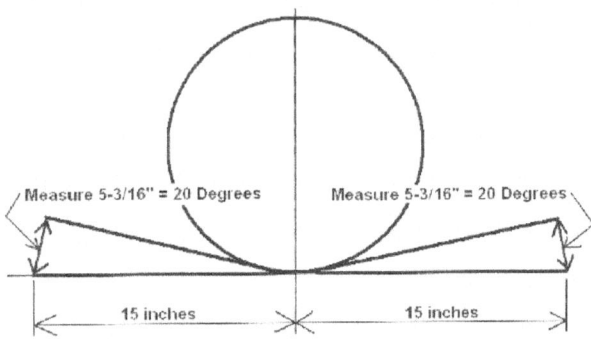

*For a digital gauge, turn the wheels to the right so that they turn twenty degrees, level the gauge, and then push the "zero" button to set zero in the display.*

3) For a bubble gauge, level the gauge and set the adjustable Caster bubble vial so that the bubble is at the zero mark on the Caster side of the tool. For a digital gauge, set zero once the wheels are turned and the tool has been leveled.

*To measure on the floor to get exactly twenty degrees of steering, cut a 30 inch piece of 1X6 wood or similar straight, flat piece and lay it against the tire. Mark mid-way on the wood (15 inches) and line that up with the hub. Mark the outside corner of each end of the wood. Turn the steering wheel until the ends have moved 5-3/16 inches and you will have turned the wheel 20 degrees.*

4) Turn the steering wheel to the left so that the right front wheel is turned past straight ahead and ends up left of straight ahead by 20 degrees.

*Turn the wheel past zero degrees on the turn plates and go past that to twenty degrees left turn. Level and read the gauge and that is the amount of caster in the RF wheel. Repeat in reverse order for the left front wheel.*

5) Again, level the gauge and then note the location of the bubble on the scale of the manual gauge and record the amount of caster in the Right Front wheel. For a digital gauge, just read the display and that is the caster amount for the RF wheel.

6) While the wheels are still turned left 20 degrees, remove the Caster gauge and place it onto the Left Front wheel hub.

7) Level the gauge and set the bubble on the Caster gauge to zero, or zero the digital gauge.

8) Turn the steering wheel to the right past straight ahead until the LF wheel has turned 20 degrees to the right of straight ahead.

9) Level the gauge and read bubble scale or display to know how much caster is in the LF wheel.

**Adjusting Caster** - To adjust the amount of Caster in each wheel, you will need to move the upper or lower ball joint fore or aft. To increase the amount of positive caster, move the upper ball joint towards the rear of the car, or the bottom ball joint towards the front. Some cars have slots cut into the upper chassis mounts or upper control arm shaft for this purpose.

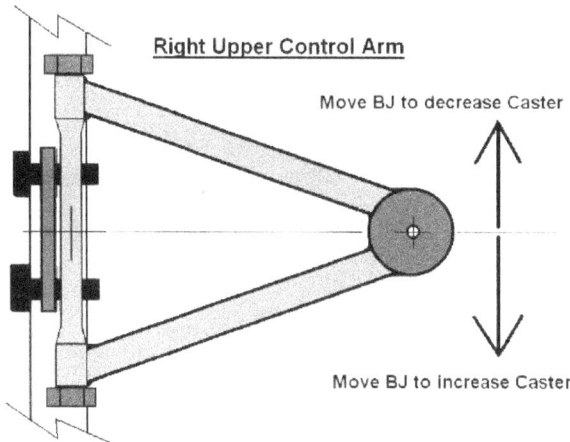

*We introduce caster into our front suspension system by moving the upper ball joint forward or backward. We can also move the lower ball joint fore or aft to accomplish the same thing. It is more common and many times easier to move only the upper ball joint.*

If you have permanently attached vertical mounting plates that the upper control arms are attached to, then you can vary the amount of shim spacing for each of the bolts that attach the control arm to the chassis. Wider spacing at the front bolt (control arm shaft inside of the mounting plate) will move the upper ball joint to the front creating less Caster at that wheel and so on. This is not the preferred method though.

*You can move camber adjustment washer shims back and forth from each mounting bolt, or slide the control arm shaft if it is slotted like this one is, to adjust the amount of caster in each wheel. This will move the upper Ball Joint forward (to reduce positive Caster) or to the rear (to add positive Caster).*

If you are not using slotted control arm shafts, once you have established the exact Caster amounts for each wheel using the above method, you should order an upper control arm that has the ball joint mount offset to give the correct amount of caster at each wheel. That way, you can use the same camber shim spacing for each mounting bolt to connect the upper control arm shaft to the chassis, which is much safer.

**How Much Caster Split** - Normal Caster splits for most short track asphalt applications are around 2 to 4 degrees of difference, left to right. The Left Front caster might be 1-2 degrees and the RF caster might be 3-5 degrees. The left front caster could also be (-) 1-2 degrees and the right camber a (+) 1-4 degrees. The higher the banking angle of the race track, the less Caster split you will need because less steering effort is needed due to the banking.

Also, the tighter the turn radius, the more Caster split that is needed. Driver preference plays a big role in getting the Caster split right for your application. Some drivers don't want any amount of caster split so they can feel the steering better.

# Exam - In The Context Of This Lesson:

### Caster Settings Are Relative To?

1) The track banking angle
2) The track turn radius
3) The type of race car
4) Driver preference
5) All of the above

### Caster Is Present When?

1) The upper ball joint is behind the lower ball joint
2) The upper ball joint is ahead of the lower ball joint
3) The upper ball joint is closer to the centerline than the lower BJ
4) 1 and 2

### To Measure Caster We Turn The Wheel?

1) Lock to lock
2) The same as when we race
3) Twenty degrees left and right
4) Until the level bubble moves

### Road Racing Cars Might Need?

1) More caster than circle track cars
2) Less caster than circle track cars
3) Zero caster split
4) The same caster on both sides
5) All of the above

### To Change Caster We Do What?

1) Move the upper ball joint forward for less positive caster
2) Move the upper ball joint rearward for more positive caster
3) Turn the steering more or less
4) 1 and 2

## Racecar Technology – Level Two
## Lesson Five – Setting Camber

The basics of changing the camber in your wheels in a AA-ram suspension is pretty straight forward and simple. You didn't take this Course to learn the basics, you took it to go beyond the basics and come to know more and gain information not readily available elsewhere.

We make camber changes by moving the upper ball joint, in most cases, in or out from the center of the chassis. This changes the angle of the wheel and tire to the racing surface. And we do that by either moving the upper control arm in or out, or lengthening or shortening the arm itself if it is designed that way. There, that is the basics. But there is much more to know and understand.

*To read your camber, we can use a simple bubble vial gauge where we read directly the camber by noting where the bubble is located on the scale. Make sure your gauge is level side to side before taking a reading. We need to note what our static cambers are as well as the dynamic cambers, or the camber when the car is at the mid-turn attitude.*

*When the camber changes, so do other settings. Here we will learn how to change cambers and how to understand what that involves. There is more than meets the eye.*

In this Lesson, we will tell you about the different ways camber changes have an effect on other chassis setup parameters, and how we can optimize camber and for what reason. Of the two most critical necessities for optimum race car performance, the first being loading on the tires, contact patch size is the other and camber dictates what our ultimate contact patch shape and size will be. Therefore, understanding all that is related to camber design becomes a very important part of chassis design and setup.

*As easier way to read cambers is to use a digital camber gauge. Again, be sure to level the gauge and then just read the number. If a negative sign shows, then the camber is negative, and if the sign is not there, it is a positive camber. We can even use a digital angle finder set against the side of the tire, but we need to be careful of the lettering and how that might influence the angle we are reading.*

**Why Change?** – The reasons why we might want to change cambers include the following. We might see where the tire temperatures are not ideal in our view for the type of car, type of tire and design of race track we are running at. We also might want to experiment with camber to see if the tire could use more or less to provide us with better traction. The tire temperatures might look good to us, but hey, what the heck, let's see if more is better.

Tire temperatures are not really the final determining factor in judging whether we have the best camber in our tires. In some suspension designs and especially tire designs, the tire can roll through camber changes such that temperatures rise across the tread, but the tire is never at the ideal camber. The temperatures lie to us sometimes. I've personally seen this happen.

An example of this is the NE pavement modifieds, being the SK division as well as the Tour modifieds. Some teams run very little right upper control arm angle, close to 5 degrees or so. With that angle, the tire runs the straightaways on the inside edge, through entry the transition on the middle of the tire and the mid-turn on the outside. When the team checks the tire temperatures, they might look OK. But the tire never has the best tire contact patch.

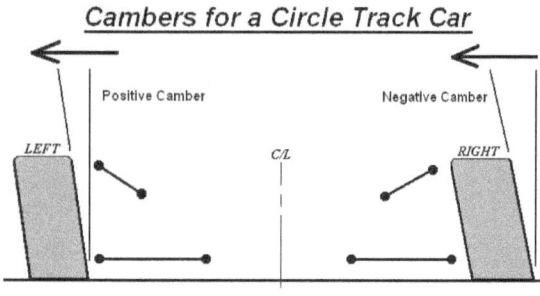

*We simply add or subtract spacers between the control arm shaft and the upper control arm mounting plate to change the cambers, right? When we do that, we are changing the chassis height at that corner and also changing the corner weights on all four corners. We need to understand how this happens and adjust the ride height to compensate.*

*Basic camber knowledge tells us that on a circle track, turning left, we need positive camber in the left front tire and negative camber in the right front tire. When the chassis travels in dive and roll, these cambers will change. How much depends on the control arm angles, especially the upper angles.*

You can get a clue this is happening by simply observing the car going through mid-turn. When your car goes out, look at the tires and see if they look like they have the camber you think they need on the race track through the turns. Visuals can tell us a lot. A straight up tire does not have the camber it needs in order to provide the most tire contact patch.

**Ride Height Changes** – When we decide we need to change the cambers, we need to know what else we are changing when we do that. A quick camber change on pit road does not just represent the "one change at a time" edict racers have followed for many years. It goes, only make one change at a time so you can better judge what that change did. Make more than one change and you will end up guessing which, or were both, the cause for the change in handling.

Changing camber will change the weight distribution on the tire you are making the camber change to as well as the other three tires. Here is how that works.

*For this design for upper control arm, we adjust the length of the links to change the camber of the wheel and tire. We need to be careful when doing this so that we don't inadvertently change the caster settings. This is not the ideal design for upper control arms by a long shot.*

On the left front, the spindle inclination, or angle of the ball joints from a front view, is usually less than at the right front. Typical left side king pin angles are 5-7 degrees for circle track cars. This puts a lot of the tire outside the intersecting line that passes through the ball joint centers.

As such, when we add positive camber to the left front, we essentially jack up the chassis. In other words, we load the LF and RR more so which unloads the RF and LR corners. This de-wedges the car (reduces cross weight) and can make the car looser, or oversteer for those who know that language.

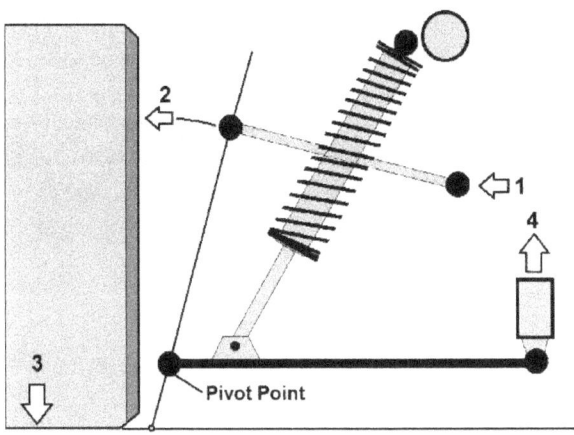

*On the left side, we can see here how the intersection of a line through the centers of the ball joints, upper and lower, intersects with the ground inside the tire contact patch. Not all suspensions are setup exactly this way, this is just a demonstration of approximately how this works. If we add positive camber by moving 1, 2 moves in the same direction. Then as 2 rotates around the Pivot Point, 3 moves down lifting 4 the chassis up. We would need to adjust the spring length to bring the chassis back to normal ride height. On the track, we cannot do this, so there is load change happening with chassis travel.*

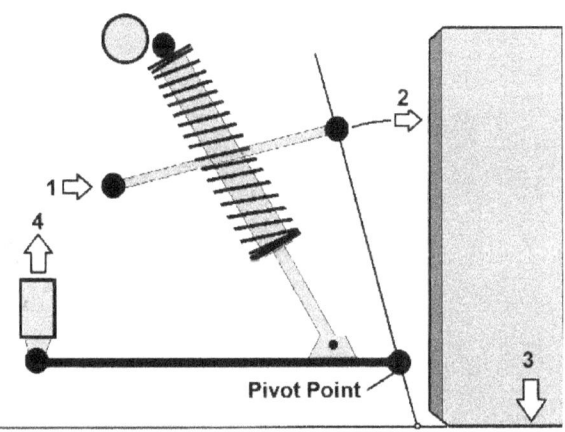

*On the right side, what we are seeing is that as we reduce negative camber, or move the wheel towards positive camber, we are also raising the chassis and adding cross weight percent if we don't adjust the spring length to compensate for the chassis height change.*

If the car turned better after the change, we might think that the added camber did the trick when in reality it was the drop in cross weight that made the car turn better by making it looser.

At the RF corner, if we reduce negative camber, we add to the wedge, or cross weight, to the car because that tire moves down with the change. Since that spindle has a higher degree of inclination, the intersecting point falls closer to the tires center of contact patch. So, the effect is less than at the LF for the same degree of camber change.

If the LF addition of positive camber de-wedges the car and reduction of negative camber at the RF adds to the wedge in the car, then they might cancel each other out, right? Not really because for each degree of change in camber, the LF changes the wedge more than the RF does. The net is still a looser car, but how much is dependent on the amount of camber change and the overall design of the suspension.

If we add positive camber to the LF and also add negative camber to the RF, we have really loosened up the car because both of those actions took wedge out of the car. The car is turning better now, but not necessarily because the camber change caused the tires to work harder.

The bottom line here is this, when you change the camber in one wheel, bring the ride height back to normal, or what it was before the change. Then you will not have changed the cross weight percent and loading on the four tires. At least not statically. Remember, make just one change at a time.

**Bump Setup Camber Change** – With the advent of the bump setups in circle track and other forms of racing, camber change of any significant amount while on the bumps is a thing of the past. Even so, there is still a lot of camber change going on from when the car is on the grid to when it is going through the turns. That camber change can cause other setup changes we might not easily see.

The tires provide better grip when the cambers are not changing and will stay very close to ideal due to very little vertical chassis movement. But what about the transition onto the bumps? What happens during that process? Few teams consider what happens then. The associated movement can cause other problems.

As we have discussed above for a more conventional car, when we change the cambers, we are also changing the corner heights and along with that, the distribution of loading on the four tires. If the chassis travels some three or three and a half inches down onto the bumps, where does the wedge, or cross weight, go to then? It has to change from where it was statically based on what we just learned.

Weight change due to the jacking effect of camber change (and that is essentially what it is, the tire jacks up, or down depending on the change) also changes the loading on the four tires.

We already know that the extreme travel associated with bump setups, from ride height to on-track height will also most times pre-load the sway bar. This adds cross weight to the setup. If the camber change from all of that travel changes the cross weight too, then where do we end up? Good question.

**Finding Loaded Cross Weight** – The major point made here in this Lesson is that camber change will change the loading on the tires and we have to know how it does that in order to plan for it in our preparation of the race car.

If we know how much the cross weight will change due to chassis travel, then we can adjust our static cross weight percent with the car at ride height. Say we gain 2% of cross weight at full travel, then all we need to do is set 2% less cross weight in the car at normal ride height from what we know we will need.

If we put the car onto the bumps with the springs removed and the sway bar preload set, we can note what the scales read. Then we can calculate the gain or loss in cross weight and therefore know how much the weight distribution changes as the car goes down.

*When a race car is down on the bumps, bump springs in this case, there will be very little vertical movement and therefore, very little camber change. Depending on the control arm angles, we could even see some change when on the bumps that we can further eliminate and we will cover that part of the design in Race Car Technology Level Three.*

To eliminate the change in weight distribution, you can adjust the bump spacing between it and the shock body to bring the cross weight percent back to what you want while on the bumps. That will compensate for the sway bar and the camber change that happens due to the high amount of travel.

One way to know the amount of change associated just with the sway bar is to push the car down onto the bumps with the bar set to neutral at ride height and see how much the bar is loaded while down on the bumps. You can then adjust your sway bar pre-load with the car at ride height so that it will not add or subtract cross weight as the car travels onto the bumps.

**Summary -** A large part of knowing your settings isn't just knowing what is happening when the car is at ride height, but more so knowing what happens when we make changes and when the car travels on the race track. This becomes a part of the art of chassis setup and one of the primary reasons why this school exists. We want you to become better at the art of race car setups.

# Exam - In The Context Of This Lesson:

## Cambers Are Mostly Set By Moving?
1) The upper control arm fore and aft
2) The upper ball joints
3) The lower ball joints
4) The lower control arm fore and aft

## Cambers Are Read Using What?
1) A bubble and vial gauge
2) A digital camber gauge
3) A digital angle finder
4) All of the above

## Why Do We Make Camber Changes?
1) To make the car turn better
2) To create a better heat pattern on the tire
3) To create a larger contact patch
4) All of the above

## Adding Positive Camber To The Left Front Will?
1) Lower that corner of the car
2) Add load to that corner of the car
3) Raise that corner of the car
4) Take away load from that corner of the car
5) 2 and 3

## Adding Negative Camber To The Right Front Will?
1) Lower that corner of the car
2) Add load to that corner of the car
3) Raise that corner of the car
4) Take away load from that corner of the car
5) 1 and 4

## Bump Setups Help The Tire Why?
1) The center of gravity is lower
2) There is better aero efficiency
3) There is very little camber change on the bumps
4) The heat on the tire is ideal

## As The Front End Travels Down On A Circle Track Late Model, The Sway Bar?
1) Provides more grip
2) Reduces the cross weight or wedge
3) Increases the cross weight or wedge
4) Provides more bite off the corner

## Lesson Six – Bump Steer

Part of the process of designing our cars for proper alignment is minimizing bump steer. Not only do we need to measure the steer as the wheel travels in a AA-arm suspension, we need to see what happens with our bump steer when the wheels are turned and what else may be affected when we make changes to our suspension components.

There is bump steer in a solid axle suspension as well, but here we are only discussing bump steer in a double A-arm, or what we might refer to as a AA-arm suspension. In Lessons 11 and 12 we will address rear solid axle bump steer.

*Bump steer is one of the most basic geometry functions in the front end of the race car. There are some interesting things to know about how bump steer occurs and how to measure it. We will also discuss ways we can interfere with the perfect bump steer settings by making chassis setup and geometry changes.*

Certain setups have evolved to where your bump steer might not be what you think it is. When you make changes to your Anti-dive or Moment Center location, you may be changing your bump steer characteristics. So, when you change to a new setup, you might be introducing BS and not know it.

The greater amount of travel associated with the soft spring setups may be out of the range in which you last checked your BS. As a result, the car may begin to behave erratically and cause the driver to be uncomfortable with the new setups because you now have excess bump steer.

We will use the term, near zero bump steer simply because most cars will never be able to have absolute zero bump steer. If you can get your bump to under 0.020 inches of toe change in the total amount of vertical travel, you will have a good design where the driver will not be able to feel the movement.

That amounts to a little over 1/64 inch of toe change. If you are inclined to want a small amount of bump in some direction for one or both front wheels, that's OK as long as you know what it is.

**Basics of Bump Steer** – As the front wheels move up and down, we want both wheels to stay pointed in the direction they were to begin with. It is most important for the wheels to have minimal bump when we are negotiating the turns. There are certain elements of the construction of the front end components that will reduce or eliminate bumps steer.

The components that make up the suspension parts for bump steer are the upper and lower control arms and the tie rods that connect the spindle to the steering mechanism, being either a rack and pinion or drag link system.

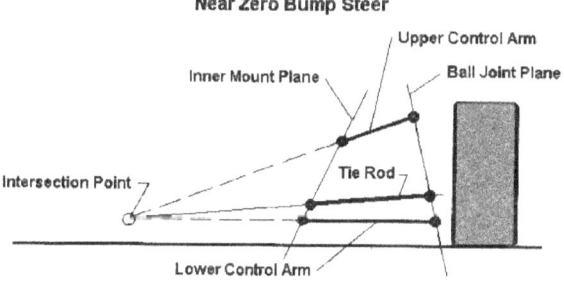

*In order for your car to have near zero bump steer, two conditions need to be met. 1) A line through the tie rod needs to point to and intersect with the Instant Center and, 2) its length need to be specific to planes created by the inner mounts and the ball joints.*

The direction of the upper and lower control arms caused by a line extending through the center of rotation of the ball joints and inner mounts of each arm, will intersect at a point we call the Instant Center (IC). This is one of the components used to determine the Moment Center location too. In order to have near zero bump steer, the intended goal, we need to setup the car so that a line through the centers of rotation of the tie rod ends points to, and intersects with, the IC formed by the control arms on that side. This is one of two criteria for near zero BS.

The second thing we need is for the tie rod to be a specific length. That length must be equal to the distance formed by the intersection of the line extending through the centers of rotation of the tie rod ends with, 1) a vertical line extending through the upper and lower ball joints, and 2) the plane that is formed by the inner chassis mounts of the upper and lower control arms.

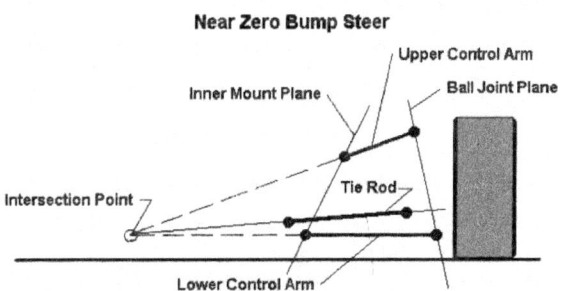

*The tie rod length requirement does not mean the tie rod needs to be positioned laterally exactly between the chassis mounting plane and the ball joint plane, it can be offset as long as it is still pointed at the Instant Center.*

This can get a little complicated because although the ball joints do form a single line, the chassis mounts form a twisted plane because of the difference in distance from the centerline of the front and rear mounts on the lower control arms, and sometimes with the upper control arms.

So, the inner tie rod intersection point is where the tie rod line intersects the plane of the inner mounts and the outer line intersection point is where it intersects the line through the ball joints. A three-dimensional geometry program can simulate this very well, but most of us don't have the luxury of owning and knowing how to operate one of those. So, we must go through the process of physically measuring the BS in our cars.

The other thing to know is that the tie rod does not need to lie within the two planes, but the length must be that length. Most of the time, the tie rod is located closer to the centerline than it would be if it were to fit inside the intersection points. That's OK because our length is still correct and we should have near zero bump steer.

**What Creates Bump Steer** – When the tie rod is not aligned with the IC and/or the length is wrong for the system, we have BS. As the wheel moves vertically, the wheel will either steer left or right. We will refer to the direction from a driver's perspective only, in this discussion.

In a front steer car, meaning the tie rods are in front of the spindle pins, if the tie rod were pointed so the tie rod line passes below the IC, then the wheel will bump in (towards the centerline of the car) as the wheel travels up. The wheel will bump out when the wheel travels down. If the tie rod line passes above the IC, then we will have bump out as the wheel travels up and bump in when the wheel travels down.

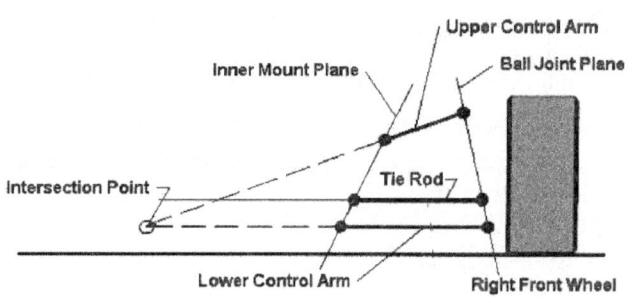

*If the tie rod is aligned to be pointing above the Instant Center, the wheel will bump out when moving up and bump in when moving down. The arc it follows will not be correct in relation to the arc of the ball joints.*

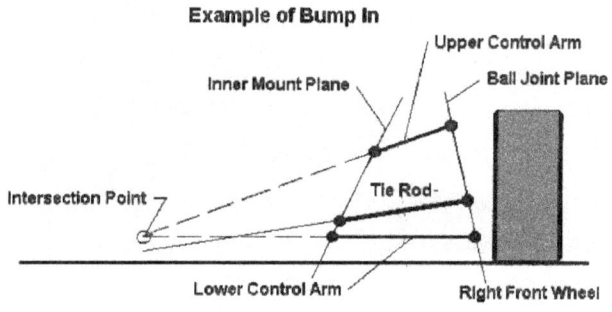

*If the tie rod is aligned where it points below the Instant Center, then the wheel will bump in when traveling up and bump out when traveling down due to the difference in arc path to the ball joints.*

If the tie rod were too short, we would have bump steer in when the wheel travels in both directions from the static ride height position. If it were too long, then the wheel would bump out as the wheel traveled in both directions from ride height.

These indicators can tell us if we have either a tie rod alignment problem or a tie rod length problem. In some cases, both may be present and that causes a very erratic motion of the wheel. To determine which of these might be present, record the amount of toe change per inch for several inches of travel in both directions from static ride height and note the tendencies. You might have perfect alignment of the tire rod to the IC and a tie rod that is wrong for length. This could be due to a poorly designed drag link or the wrong width rack and pinion steering unit.

If the tie rod is the correct length, and the wheel bumps in or out, we will know how to reposition the ends of the tie rod to make it intersect the IC based on what we discussed above.

**Design Changes That Affect Bump Steer** – We could buy a car that was near perfect for BS and then make design changes that would unknowingly change our BS. When racers and manufacturers started to install spindles that were designed for rack and pinion systems into a drag link system some years ago, they inadvertently changed the bump steer characteristics, along with the Ackermann geometry.

With the drag link system, the outer tie rod end should be closer to the centerline of the car than the lower ball joint by using angled (from a top view) steering arms. This design feature cancelled out the natural tendency for the system to cause toe changes as the car was steered.

When the new, "rack" spindles were installed, with their straight ahead steering arms, the length of the tie rods changed necessarily. This created a bump out situation as the wheel traveled up and down. New drag links with the inner tie rod ends placed further out were needed so that the tie rod would remain the correct length to eliminate the adverse BS those spindles created.

If we make changes to the front end geometry to improve our Moment Center location, we might accidentally change the BS characteristics at the same time. For example, when we install extended lower ball joints to take angle out of the lower control arms, we change the angle of the lower arm and move the IC height. The tie rod may not now intersect with the IC and we have now introduced BS. When making changes to the arm angles, we need to re-align the tie rod so that it stays pointed towards the IC.

*In this example, the race team added spacers to the lower ball joint to change the angle of the lower control arm for better geometry. The outer end of the tie rod shown closest to you has not been moved and is no longer pointed at the IC. This system probably now has a considerable amount of bump steer. Remember that when you make changes to the geometry of your AA-arm suspension, you need to re-check your bump steer and make corrections.*

**Anti's Affect Bump Steer** – In both dirt and asphalt racing, anti and pro-dive is used in various degrees. If we make changes to enhance those effects, we could be causing changes to our BS. This is because with anti-dive for example, when the right wheel travels up, the upper ball joint moves towards the rear of the car and this rotates the spindle from a side view, counterclockwise. This rotation moves the outer tie rod end upwards and changes the angle of the tie rod. Now it no longer points towards the IC.

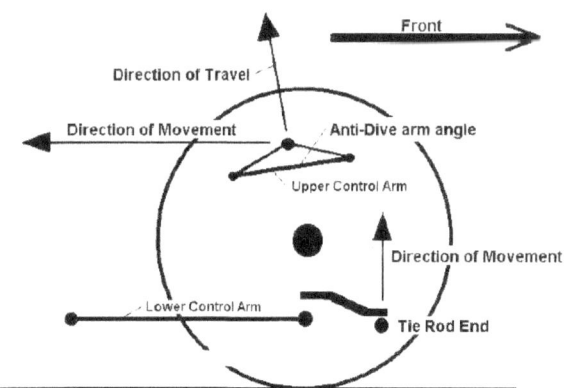

*With Anti-dive, as the wheel moves up, the upper ball joint moves rearward, and in some cases the lower ball joint moves to the front. This causes the spindle to rotate and with it the steering arm. The outer tie rod end will move up changing the front view tie rod angle. This changes the alignment with the Instant Center and will change the bump steer characteristics.*

Where we had near zero BS before with no anti-dive, we now have BS when we introduce anti-dive. With pro-dive, we see a similar affect, the tie rod end moves down with vertical travel and again the tie rod is miss-aligned with the IC. If you originally checked your BS and found it acceptable and then experimented with Anti's, and didn't recheck your BS, you could, and probably do, have a problem.

**Steering Affects Bump Steer** – When we steer our front wheels, we change the angles of our tie rods due to caster, camber and degree of spindle on both sides. The tie rod ends travel in an arc that is not parallel to the ground.

This changes the outer tie rod end height and therefore the BS. It is for this reason that we recommend doing your BS with the wheels both straight ahead and then again with the wheels turned equal to the mid-turn steering angle you steer to at the track you will run.

**Measuring Bump Steer, Some Tips** – We can measure our BS using several different types of equipment. There is the double dial caliper system, the single dial system and the laser system. Each one will tell us if the wheels steer when they are moved in bump (moving up) or in rebound (moving down).

The most common tool is the bump steer gauge. It consists of a flat plate bolted to the hub and a stand that holds dial indicators. It comes in two configurations, the double dial indicator type with a stationary stand and the single dial type with a swinging stand. With the single dial type, when the wheel moves vertically, the stand follows the plate. With the double dial type, the stand is stationary and the two dial pins are always moving one way or the other.

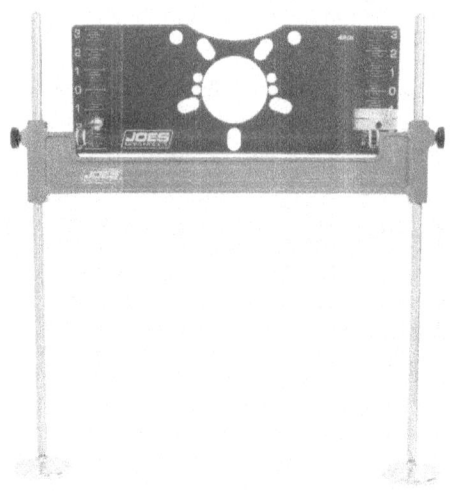

*This is a bump steer gauge that uses one dial indicator. It moves in and out with the wheel movement and uses a roller follower on one side and a dial indicator on the other. If the wheel steers when moved vertically, then the dial will move in or out, depending on the way the wheel is steering.*

*The most common way to measure bump steer is to support the car on jack stands and using a jack, move the spindle up and down in the range of motion it will have during a racing lap. If your wheel travel is 3-4 inches, moving the wheel one or two inches up and down will not simulate the conditions. For most asphalt circle track racing, both front wheels travel up in relation to the chassis. For dirt cars, the left front may well travel down considerably.*

*Using a laser system mounted to the wheel hub, you need to place targets in front of and to the rear of the wheel at equal spacing. Preferably, measure the tire diameter and multiply by 3, 4 or 5 so your multiplier used to divide into the readings is an even number. Example: for a 28 inch diameter tire, go 84, 112 or 140 inches out to each target. The farther you go, the more accurate the readings will be.*

*You can check your bump steer with a dedicated gauge available from many suppliers, or a laser system like this one from DRP Performance Products. This one uses a remote laser that projects a laser line onto a fixture mounted to the wheel hub. Whichever one you choose, make sure you understand the process and how bump changes when other components are adjusted.*

If the double dial system has zero BS, then both dials will move together the same amount. If the front dial moves farther as the wheel moves vertically, then we have bump-in at that wheel. If it moves less, then we have bump-out. Be sure to count the number of turns each dial makes when moving the spindle vertically. Subtract the readings to find the BS amount related to the distance the wheel has moved. We usually refer to BS as decimal inches of bump per whole inches of travel.

The single dial indicator gauge is a little different and one I personally like. Using a swing stand, it has one dial that rides on one side of the plate and a roller that rides on the other side. As the wheel moves vertically, the stand follows it in and out. If the wheel has zero BS, then the roller and the dial shaft will move together the same amount and the dial will not change its reading.

If the dial does move, it is recording the total amount of bump steer. Since the roller is stationary, the dial records the movement between itself and the plate, or the total distance between them. With the double dial, when you subtract the two readings, you have the total amount of bump steer.

At the right front, if the dial is on the front of the plate, movement causing a greater number means the wheel is bumping out. If the reading is getting smaller, or the dial shaft is moving out, then we have bump in.

We can also use a laser alignment system to check for bump steer. When using a laser systems, the laser is mounted on the hub or wheel and a targets is placed ahead and behind the wheel center the same distance. This way, any difference in movement of the laser on the two targets, as the wheel moves vertically, will be divided by the distance to the target from the center of the wheel divided by the diameter of the tire.

So, if a tire where 28 inches in diameter (an 88 inch circumference tire), and the targets were 112 inches away, we would divide the difference in movement of the laser front to rear by four (112 / 28 = 4). If the wheel were bumping 0.030 inch, the differential readings on the targets would be 0.120 inch or about an eighth of an inch.

Remember too that when we move the wheel up or down, it will most times move in towards the centerline of the car, or move out. That is why we take a difference in movement on the targets, not the movement itself of the laser lines. If the both move the same distance, then we have near zero bump.

**Gauge Ratio** – It is important to consider that most bump steer gauges are not the width of the tire diameter for practical purposes. So, we must translate the readings from the width of the tool to the tire diameter if we want to record the bump as being equal to toe.

*We can see how the width of the dial indicators is much less than the tire diameter we use for measuring toe. To translate the numbers for Bump into toe, we need to multiply the results by the ratio of the gauge width to the tire diameter. Divide the gauge width into the wheel diameter. For example, 0.030 inch of change on the bump steer gauge could be as much as 0.050 inch of toe change.*

This is easy, just divide the measurement between the dials, or the dial and the roller depending on the type, into the tire diameter and use that number to multiply times the bump reading to see what the bump is in toe equivalent, which is something we all more readily understand.

**How To Measure Bump Steer** - With the car at ride height, measure your shock lengths and write those dimensions down. We want to reference those measurements when we move the wheel in the range of motion it will experience on the race track.

Now you can put the car on jack stands at a height closer to ride height but not too high and remove the front wheels. Remove the coil over, or shock and spring in a "stock" spring car. Check the suspension for freedom of movement by moving it up and down. If the motion feels rough or sticks at points in the travel, you might have dry ball joints or corroded bushings at the chassis mounts. Correct these problems before you proceed.

Start with the right front wheel. Place a jack under the lower control arm and bring the wheel up to your ride height shock length. This is where we will begin to take measurements. For this exercise, we are using a standard bump steer gauge setup that uses a plate mounted to the hub and either one or two dial gauges.

Point the wheels straight ahead and mount your bump steer plate onto the wheel and rotate it so that it is level with the floor and then move your floor fixture up to the plate. The dial indicators will move away, or out, so be sure to bring the fixture towards the wheel enough so that the indicators can move out without bottoming out.

Note the position of the dial shafts on the plate. We will be noting how far the wheel moves vertically and recording measurements at one inch increments as we go along. So, move the dials to the zero line or some line on the plate. Now zero each dial.

Jack the wheel up one inch slowly noting how far each dial moves in relation to each other in a two dial system. Each dial will go out and the indicator will go backwards. So, you will need to be aware of the distance the dials move. If one dial is outrunning the other considerably, things are really bad in the bump steer department.

If the rear (one closer to the rear of the car) moves out faster than the front one, then you have bump out. The reverse is true for bump in. If the movement is more than fifty or sixty thousandths in that first inch, you need to stop and look for the problem.

A wheel that bumps out a lot either has a longer than necessary tie rod or the angle of the tie rod is wrong and the outer end is low and/or the inner end is high. Remember that the tie rod must point to the instant center created by extending the planes of the upper and lower control arms. Bump in is the reverse of the above.

If you can look at the tie rods from a front view looking roughly parallel to the ground plane, you can visualize where the tie rod is pointed and where the instant center probably is located. If it looks like the tie rod will never intersect the IC, start making changes to the height of the ends to get it pointed correctly. Once it looks like it is more so inline, we can proceed.

Let's assume we are getting little difference in movement of the two dials. Record the dial readings at the first inch of travel. Then jack the wheel another inch and record those readings. Now note how much the two differ in total movement. This is your average bump steer per inch of wheel travel in two inches. We are ideally looking for numbers under 10 thousandths of an inch.

Many modern race cars travel the front wheels much more than in the past. For the right front wheel on dirt and asphalt, it is now normal to see three to four inches of travel. So, we need to now bump our wheel up another two inches for those cars, recording one inch at a time.

Your readings may indicate many things when going this far with bump. What you had for bump in the first two inches may move back toward zero bump as you go higher, or it may continue to bump in the same direction even more. If you have less than 20 thousandths in the first two inches and it goes back towards zero, you are probably good to go.

If the bump steer continues to increase, note which way it is going. The back dial moving farther out and reading less than the front dial indicates bump out. Again, adjust your tie rod end heights moving the inner mount down and/or the outer mount up. Again, the reverse is true for bump in.

For readings on the left with a two dial system, the same holds true as to the movement of the front and rear dials. The rear dial moving out more and reading less equals bump out. We need to think out the direction of the movement of the wheel for our car. For dirt cars, the left wheel moves vertically up and down. So, we typically bump that wheel up and down two inches each way to read bump steer.

For asphalt cars, the left front will, or should, always move up in bump. For bump setups, that movement might nearly equal the right front movement upwards of 3 to 4 inches. So, take into account your type of setup and how much each front wheel moves when checking your bump steer.

For single dial bump steer gauges, you will only be reading the one dial. The upright follows the plate as the wheel moves in towards the centerline of the car when it is bumped and moving upwards. This is admittedly easier to use because you only need to record the movement of one dial.

In bump, moving up, on the right side, with zero bump steer, the dial should not move at all. With the dial mounted on the floor fixture towards the rear of the car and the roller to the front, if the dial is moving out and the reading is showing less, then you have bump out. The front of the tire is moving out and the back is moving in.

For bump in, the reverse is true, the dial will indicate an increase in reading since the shaft is moving in. The reading on the dial for each inch of vertical movement is the bump steer amount.

When checking the left wheel, the gauge will be at the front and the roller at the rear. Now your movement for bump out is reversed, and the readings will increase and the dial will move in. The reverse is true for bump in.

**Summary** – Once we understand all of the things that affect bump steer, we will know when we need to re-measure the car after a chassis or setup change so we can maintain near zero bump steer. If you make Moment Center changes or Anti's changes or spindle changes, or even change setups from conventional to bump, re-measure your BS.

If you have only measured bump at static ride height with the wheel pointed straight ahead, maybe it's time to re-measure the bump at mid-turn configuration. Correcting bump steer when the wheels are turned could make a difference in how the car feels to the driver as the car moves vertically on corner entry and when going over those ruts on a rough dirt track.

# Exam - In The Context Of This Lesson:

## Bump Steer Happens With/When?
1) Turning the steering wheel
2) Vertical movement of the chassis causes the wheels to steer
3) Steering away from bumps on the race track
4) When we make changes to geometry or setups
5) 2 and 4

## Bump Steer Can Occur When?
1) The tie rod is not pointed at the Instant Center
2) The tie rod is too short
3) The tie rod is too long
4) All of the above

## Anti's Affect Bump Steer By?
1) Changing the control arm angles
2) Changing upper control arm angle
3) Moving the outer end of the tie rod vertically
4) Restricting movement of the wheel

## With Double Dial Bump Steer Gauge?
1) Both dials move together
2) Only one dial moves
3) We can read the bump amount using one dial
4) The readings equal the toe steer amount

## With Single Dial Bump Steer Gauge?
1) Both the dial and roller move together
2) We only read the one dial
3) The readings do not equal the toe steer amount
4) All of the above

## Racecar Technology – Level Two
## Lesson Seven – Ackermann Settings

Ackermann, as described in RCT Level One, is a very destructive effect in race cars. We need very little of the Ackermann effect as a part of our steering system. In this Lesson, we will tell you how to check for Ackermann and how to adjust your system to eliminate excess Ackermann.

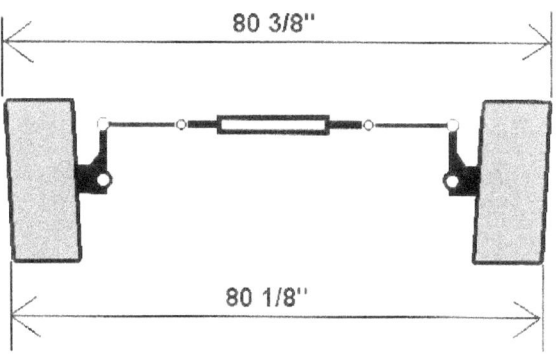

*Toe is when the distance between the front edge of the tire is more or less than the rear edge. Toe-in is when the front is less, and toe-out is when the front is more than the rear. This car has ¼" of toe out. For short track racing, we almost always use toe-out. Ackermann and Reverse Ackermann are caused when we turn the steering wheel and the system gains or loses toe. For most types of racing, we need very little or zero toe change, or Ackermann, when we steer the car.*

*You can easily check your steering system for Ackermann. Use a string method or laser system. In this Lesson we will tell you how to measure and how to fix excess Ackermann.*

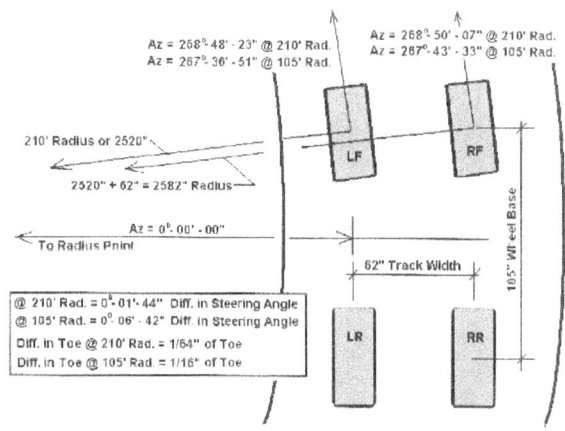

*This chart represents calculations done with a very sophisticated coordinate geometry software program. We see where we need very little difference on direction of the left wheel verses the right wheel in order for the tires to track along their individual radii. We are talking about tenths of degrees of difference, not whole degrees.*

When we have Ackermann effect present in our steering design, it means that the amount of toe-out increases as the steering wheel is turned. With Reverse Ackermann, the toe is reduced. There are different static settings for front end toe that are dependent on the radius of the turns, the banking angle, and the type of tire used.

**Checking For Ackermann** – To check to see if, and how much, Ackermann you have, you can use either simple string or if you have it, a laser system. The process is the same, just the tools are different.

Aim your wheels straight ahead in the shop, or anywhere for that matter. String along the sidewall of your tire making sure not to go over any lettering or other extrusions. Go out ten feet in front of the axle and place a piece of tape on the floor under the string line.

You will carefully lay the string so that it just touches the edge of the sidewall in the front part of the tire while having someone hold the string wrapped around the side of the rear part of the sidewall. Bring the string down to the tape and make a mark with a felt tip pen.

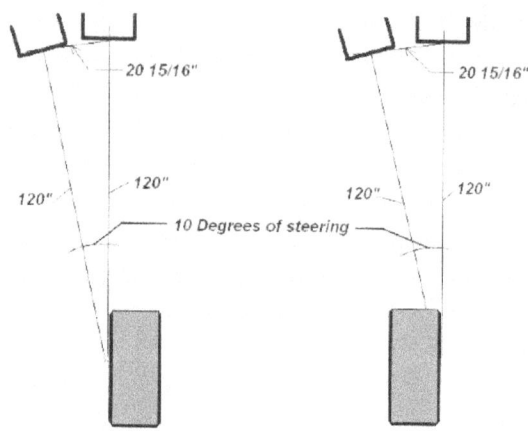

*The easiest method for checking for Ackermann is by use of strings pulled across the tire sidewalls or a laser system attached to the wheel hub. With strings, we simply pull the string along the sides of the tire bulge where there are no lettering or other protrusions. We make marks out ten feet from the axle at straight ahead and again when the wheels are turned. The difference in gap between the marks should be zero if you have aero Ackermann.*

Go to the other side of the car and repeat the above for the left front tire. When you have two marks, one for each side, turn the steering wheel to the same amount as you would at the track going through mid-turn. Then repeat the string method for both sides.

For road racing cars, you should turn the wheel both ways the same amount to check for Ackermann in both steering directions. You system might not be symmetric, especially if your car is a converted circle track car.

Now you have two sets of marks, one set for the right wheel and one for the left. Measure the distance between the straight ahead mark and the "turned" mark for each wheel. If you measure the same on each side, you have zero Ackermann. The direction of the tires is exactly the same when the wheel are turned as when they are straight ahead.

If you measure a half inch more for the left wheel than the right wheel, you have gained about one 1/8 inch of toe over what you started with. This is the most you'll ever need and might be good for very short ¼ mile ovals. For longer, faster tracks, you'll want less.

When using a laser, you will do the very same procedure, except you'll be using the laser instead of a string. Attach the laser to the hub and point it at a target ten feet in front of the axle, or down at tape on the floor. You'll make marks and measure between them.

If you are measuring a race car that turns left and right, do this same procedure with the wheels turned right also. These cars could include dirt cars and road racing cars. You might be surprised to learn that the Ackermann may be different each way as you turn the wheel.

**Correcting For Ackermann** – If you find that you have too much Ackermann, or if you have Reverse Ackermann where you lose toe (i.e. the left measurement is less than the right side measurement) and you want to correct it, here are some methods for doing that.

If there is excess angle in the tie rods from a top view (the inner mounts rearward of the outer mounts) there will be Ackermann present. Reducing this angle reduces the Ackermann and fixes it for turning both left and right, like you would do in a road racing car or a dirt car on an oval track.

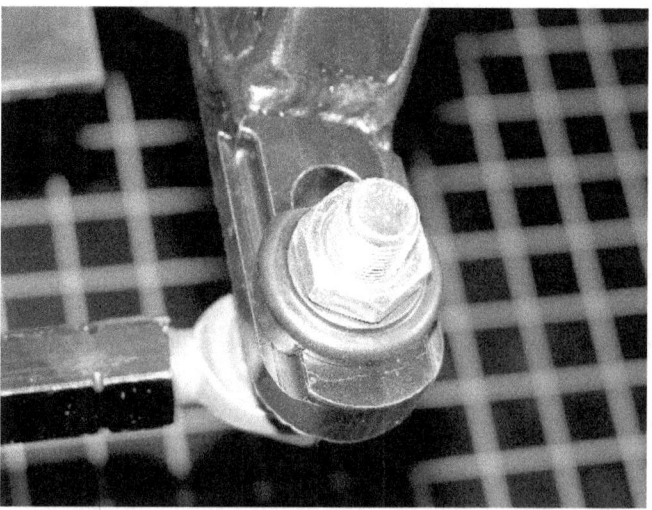

*Most modern race cars that are designed with the rack and pinion steering systems use slotted steering arms so that teams can adjust the Ackermann in their cars. For most applications, the tie rod end bolt will be to the very front of this slot. Be sure to check your Ackermann using the methods described to be sure of what you have.*

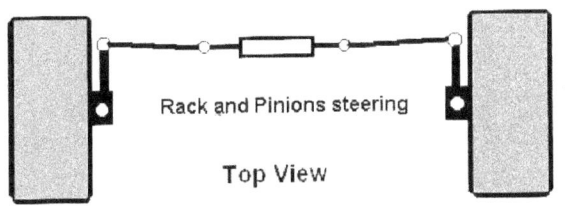

*The rack and pinion steering systems can be adjusted for Ackermann by moving the rack (at the center) forward or rearward to change the top view angle of the tie rod. The greater the tie rod angle, the more Ackermann we have in the system.*

For rack and pinion designed steering, one way to adjust for Ackermann is to move the rack forward or aft. To reduce Ackermann, move it forward and to increase it, move it rearward as in the case of Reverse Ackermann. Make sure both steering arms are the same length when making this change.

Another way to adjust Ackermann in a rack system is to change the length of both steering arms at the spindle. Be sure to change both arms the same amount and make sure they are the same length.

For drag link steering systems, we suggest a similar fix. Instead of moving the rack, we move the part of the drag link where the tie rods are attached forward if there is too much top view tie rod angle. This is not easy to do and usually involves installing a new drag link designed to place the inner tie rod mounts more forward. As with the rack system, we can (if possible in the design) change the length of the steering arms, moving both the same amount.

Some teams have taken it upon themselves to re-design the drag link, but we don't recommend bending or cutting and welding steering parts or doing something that might be unsafe. It is up to you to make sure any changes you make to your car are safe.

In past years, race cars fitted with the drag link steering systems used spindles that had steering arms with two holes. This was intended to be used to adjust the speed, or quickness, of steering by either using the front holes, or the back ones. Teams that experimented by using the back hole on the left and the front hole on the right learned very quickly that this produces a large amount of Ackermann and the car refused to turn after that change. Please don't do this. Small changes in the length of the steering arms have a pronounced affect.

Again, the above fixes are perfect for road racing cars and dirt cars where you would be steering in both directions and need to have the same toe settings whichever way you steer. But for race cars that only turn the steering wheel left, there is an easier way to go about this.

For a circle track car that always turns left, we can usually lengthen the left steering arm and that reduces the Ackermann effect. For stock type of cars with cast spindles, you'll need to cut and add material and weld back together. We would never recommend doing something that would be unsafe and you will need to determine if any changes you make are safe.

For most modern late model cars, the spindles are made with slotted steering arms, sometimes on both sides. You could lengthen the left arm and/or shorten the right arm length to reduce Ackermann. Remember that small changes in the length of the steering arms make a big change in Ackermann.

**Summary** – Ackermann was a design feature for early automobiles. It helped the wheels to track along the turning arcs in very tight turns having very short radii so they didn't create ruts in gravel driveways. In most types of racing, we have much larger radii for our turns and therefore, we don't need the Ackermann effect.

We have shown through calculations that a very small amount of Ackermann, less than an $1/8^{th}$ inch for most tight turns, is all we need. This represents a fraction of a degree of steering, not whole degrees as some race teams think.

# Exam - In The Context Of This Lesson:

**Ackermann Is Present When?**

1) The wheel steers when it bumps
2) Only one wheel turns
3) We gain or lose toe when the wheels are turned
4) Our steering arms are too short

**We Can Correct and Reduce Ackermann By?**

1) Moving the steering rack fore or aft
2) Changing the lengths of the steering arms
3) Moving the drag link forward or backwards
4) All of the above

**For Cars That Turn Both Ways, We Reduce Ackermann How?**

1) Shortening the left steering arm
2) Lengthening the right steering arm
3) Make both steering arms the same length
4) Moving the rack or drag length fore or aft

## Lesson Eight – Alignment Front To Rear

In this Lesson, we will explain the concept of the Alignment of all four wheels and how to accomplish that. This Lesson will involve the initial alignment at static ride height and not dynamic alignment as the chassis moves which is a subject we will speak about later on in RCT Level Two. The overall thought presented here is that the four wheels need to be pointed in the same direction and lined up relative to the type of racing you do.

*At this point in the RCT Course, Level Two, we need to align our wheels and tire contact patches. Many handling problems can be traced back to poor alignment issues. In this Lesson, we will tell you how to align your wheels and in what order using different methods.*

Of all of the setup parameters, including moment center location, alignment ranks at the very top of the list. Why? Because when your alignment is wrong, no other setup parameters or changes will be able to compensate for it. The alignment package has a tremendous influence on the way a race car handles.

Alignment is a first priority in setting up any race car. It really doesn't take long and the benefits are as great, or greater, than any other setup task you will perform. In this Lesson we will explain race car alignment, why it is necessary and how to do it. For this exercise, we will be aligning a race car with a AA-arm front suspension with a solid axle rear suspension.

For formula race cars that are designed with the AA-arm front and rear suspension, the alignment will be for toe only since those cars are symmetrical and the tires and wheels are equidistant from the centerline.

The elements of proper alignment of the subject race car are as follows:

**Toe Settings** – For the purpose of this Lesson, we will set the car with zero toe, meaning that the front, and rear tires are pointing straight ahead. This will facilitate some of the procedures we will undertake. We will provide a separate Lesson on proper racing Toe settings later on in this Course.

**Front to Rear Tracking** – The tire contact patches must track along the same path and parallel to the centerline of the car for all race cars. In most cases for circle track racing where the cars turn left, we need to line up the right side tire contact patches. Our final alignment will show the right side patches are in line with one another along with the rear end, or rear axle, being perpendicular to that line.

For road racing cars and formula type cars, we will place the wheels equal distance off the centerline of the chassis for the front and rear wheels and point them parallel to the centerline for alignment purposes. In many cases, the track width (distance between the centers of the tires at the contact patch for opposing wheels) is different front to rear. We will set the toe for both ends of the car later on after the alignment has been completed.

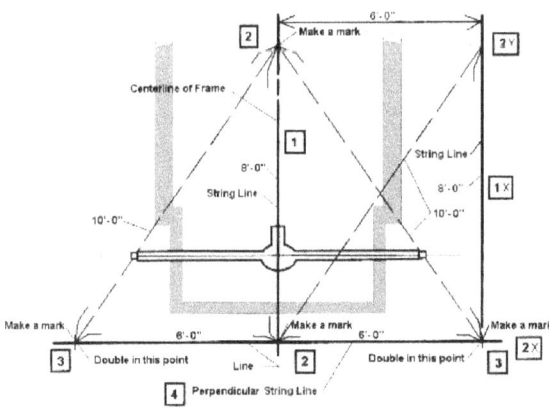

*In the past it was common to string a car by making marks on the floor. This is a good method for checking rear end square and can be used to line up the right side tires for circle track cars. It is a little slower to do than the raised string method but can be more accurate due to the process of dropping points down to the floor with a plumb bob and then making direct measurements to the rear end. This sketch shows how to setup a parallel line outside the chassis as well as a line that is ninety degrees off of centerline to assist in rear alignment.*

**Rear Alignment** – The direction, in relation to the chassis, that the solid axle rear end, or each of the wheels in a AA-arm rear suspension, are pointed can dictate how a car will behave in the turns. For example, on turn entry, if the rear wheels are pointed to the outside of the turn from the chassis centerline, no amount of setup tuning will prevent the car from being loose, or oversteering in road racing terminology.

That looseness will stay with the car throughout the turn, especially ruining the turn exit. If the rear end is pointed to the inside of the turn from the centerline of the chassis, the car will be tight, or understeer, through the middle and off the corner.

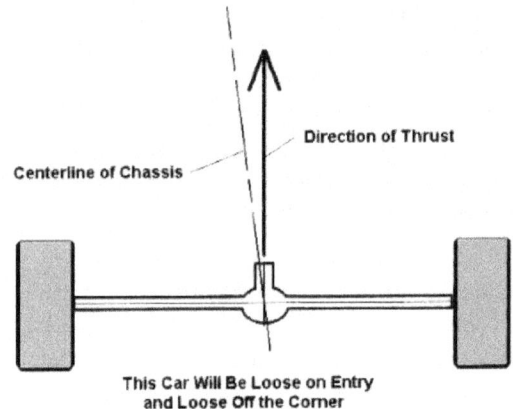

*In this example, the rear end is pointed to the right of the centerline of the car. This will help free up a tight car in a left turn, but will produce a severe loose condition under throttle application off that same turn. The loss of bite off the corner will hurt the lap times more so than what might be gained from freeing the car up for corner entry, which is usually the reason why teams do this.*

There should never be any reason to miss-align the rear wheels at static ride height. They should always be perpendicular to the centerline of the chassis and centered on the centerline of the chassis for road racing and formula cars, and with the outside tire contact patches lined up for circle track racing. This does not include rear steer, a subject we will get into later on in this Level Two course.

**Measuring The Car** - There used to be only one reliable way to align a race car and that was by using a string and either measuring to the tires at hub height or at the floor by creating right triangles on the floor to measure from. That is still a viable way to do it and necessary for the lower budget race teams.

A quicker and more accurate way to align the car in all areas is by the use of a laser system. The key to maintaining accuracy in a laser system is to be able to check the tool to make sure the beam is truly tracking parallel to the wheel and perpendicular to the mounting device. The actual laser must also be tested to be sure it is pointing inline with its base structure.

*The True Track Laser System sends a single laser beam to the ground and to targets placed at the front and rear of the car. The system can check wheel base, alignment, Ackermann, toe, etc. It can also be checked for accuracy because we always need to check to see if the beam is pointed parallel to the body of the laser. We do this by finding the direction of the beam by its showing on a plate ahead of the car. Then we remove the laser, turn it 180 degrees and remount it to the hub fixture. We then rotate it around to again point to the target to see if it again hits the same spot on the target. If not, the beam is not pointing parallel to the laser body. This means we cannot rely on the laser for accuracy in doing any kind of alignment.*

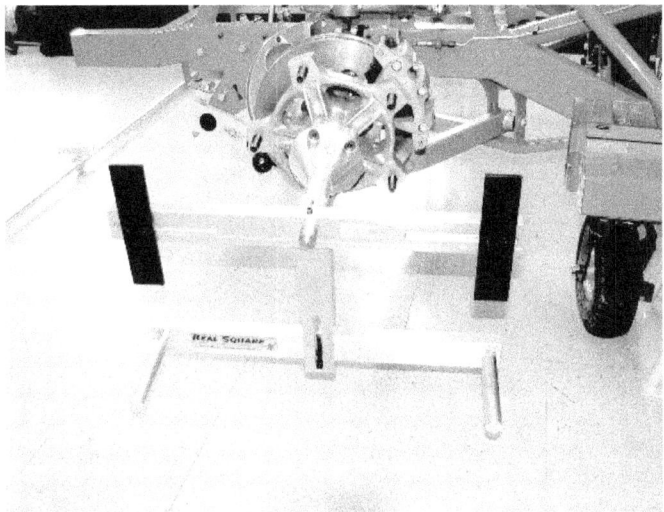

*The Real Square™ Alignment Systems by DRP Performance uses scaled bars attached to a fixture that bolts directly to the hub. After the lasers are setup and aligned with the chassis, the wheel alignment and toe can be read on the scaled bars. Ackermann can also be checked with this system along with other parameters. This system too must be checked for accuracy of the direction the lasers are pointed. When the lasers are mounted to a bar, the measurement between the laser beam can be checked at different distances from the car. If the measurements are the same, then the beams are parallel and we can rely on the system.*

The units I have used in the past are the True Laser Track, Gale Force Laser and the Real Square™ Alignment Systems. Each of these systems allow for checking the accuracy of the laser beams. This must be done each and every time we use the tool to check alignment.

The laser systems attach to the wheel hubs (models for both Wide 5 and 5 on 5 hubs) and are pointed at targets or measuring scales. For the "Analog" method, the nylon string we used was from the local hardware store, or you can use heavy monofilament fishing line.

We will go through the process using both methods so you can see how to perform each step. Note that the order in which we will proceed through the alignment process is important. Performing your alignment out of order can upset portions that have already been done.

**Step 1 - Wheel Run-out Check** – Check both the front wheels and the rear wheels for Run-out. This means that as the wheel rotates, the outer edge of the tire will wobble slightly. We must compensate for this slight distortion by finding the extreme high spot at a point equal in height to the hub height. We can simply use a jack stand to hold the tape steady and rotate the tire noting the distance from the stand. Once we locate the high point, we mark it with an arrow and then rotate the tire (be it front or rear) so the arrow is at the top pointing straight up.

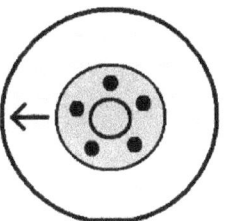 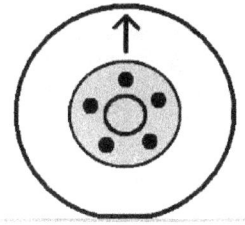

1) Mark the high point at axle height with an arrow.

2) Rotate the tire so that the arrow is pointed straight up.

*You need to check for run-out in the wheels before you start your alignment process. We tell you how in the Lesson. Move the high point to the top when you are finished. Now, check the toe at the rear end for all suspensions. Use toe plates or toe bars for the "analog" method and the laser systems for a more accurate assessment. Even small amounts of toe-in or toe-out are not acceptable when doing alignment. So, you might need to straighten any straight axle rear end before proceeding. When checking toe, be careful how each person holds the plates so that the measurements are consistent. Do the measurement several times to ensure accuracy.*

When using the laser systems, attach the laser to the hubs remembering to thoroughly clean the surface of the hub and make sure there are no protruding threads from the bolt holes. It is recommended that you go over the hub surface with a flat file to eliminate any bulges or protruding edges of metal that would cause the laser to not be aligned properly.

Follow the manufacturer's recommendations for setting up the laser systems. Remember that the accuracy of the measurement is directly related to how closely you follow the directions and how carefully you read the system. There is a logical progression to alignment and each company has put a lot of thought into the methods. The end results, if properly applied, will be the same.

**Step 2 - System to Frame Setup** – Once the rear wheels have been checked for zero toe with both pointing straight ahead, square the laser system to the frame making the laser line parallel to the centerline of the car. Most car builders will align at least one frame rail parallel to the intended centerline of the chassis for circle track cars. It may be located on the right side or as the weight box on the left side next to the driver.

*The offset distance of the laser beam to the frame can be taken by using a laser or by the use of a string. The string is ideal for a low budget team, but the laser systems are much more accurate and do more alignment tasks in a shorter amount of time.*

You can also align the laser system to be parallel to the centers between the front and rear clip rails closest to the main frame rails for perimeter and road racing cars, or those where the dimensions from the centerline to the wheels are equal for both sides of the car. Some circle track cars are built this way and all road racing and formula cars are built this way.

For the analog method, this setup can also be done using strings. We will setup a "box" with strings on each side and at the front and rear. We use plumb bobs to make marks on the floor off the outsides of the frame rails or the centers of the front and rear suspensions. For Perimeter cars, or ones with no offset in the chassis, we can split the measurement between the frame rails to find the centerline of the chassis.

For offset chassis, we should be able to use a straight rail, or we could again split the front and rear frame rails. Measuring between the marks on each side, we can split the distance in half and place a mark at the halfway point which will represent the centerline of the car.

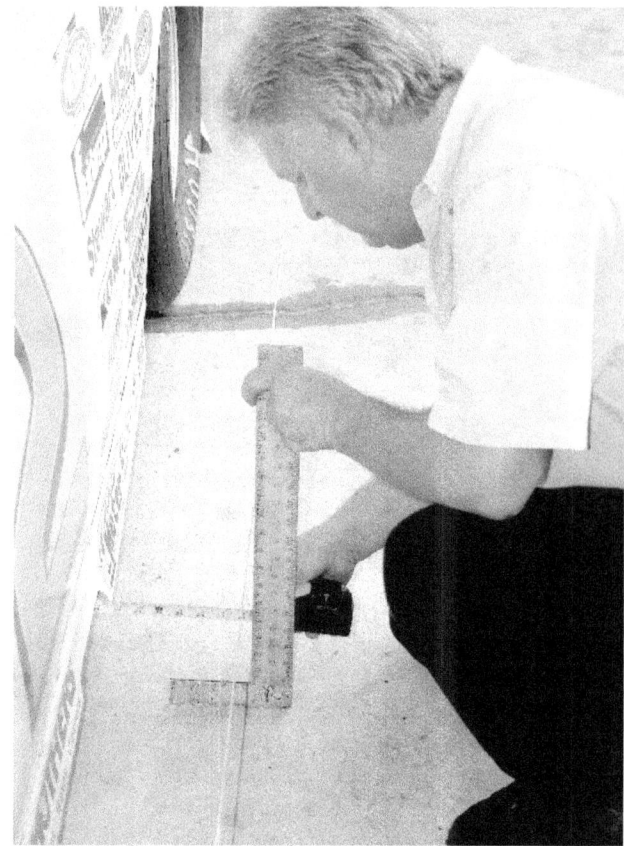

*When using the string method, make sure your tape is level and not bent or held at an angle other than level. A tight string is always straight, but we can introduce errors in our measurements if we are not careful. Try not to eye ball too high above the string or your reading of the measurement will be in error.*

**Step 3 - Center the Steering Box** – Center the front steering system. This is done by turning the steering wheel lock to lock and back half the number of turns from full lock in either direction. Once mid-rack (or mid-box with drag link steering) is found, lock the steering shaft against the frame rail with two vise-grip type of pliers applied in opposite directions. This will become obvious when you try to do it.

We want to make sure the steering is centered and then we can adjust the wheels to be pointed straight ahead. Once the steering box has been set to center, adjust each tie rod length so that the right and left wheels are pointing straight ahead. With the laser systems, this can be done quickly and accurately.

With the string method, run a string down each side of the car at hub height and parallel to the centerline you have established. Then, simply measure to the side walls of each front tire, or to the rim, with the tape measure and make both front and rear measurements equal by adjusting the tie rod lengths. We can set our race toe later on after we have aligned the car.

**Step 4 - Rear Tire Contact Patch Alignment** – Once the front wheels have been adjusted to point straight ahead and parallel to the centerline, we need to align the rear tire contact patches to the front tire contact patches. We do this by using our laser systems as described in the user manuals for each system.

If using hub mounted lasers, mount the laser to the front wheels only at this time. When moving the rear end to adjust the rear wheels to proper placement of the tire contact patches off the centerline, the rear end may steer. If the lasers were mounted to the rear wheels, then they would change direction as we move the rear wheels laterally.

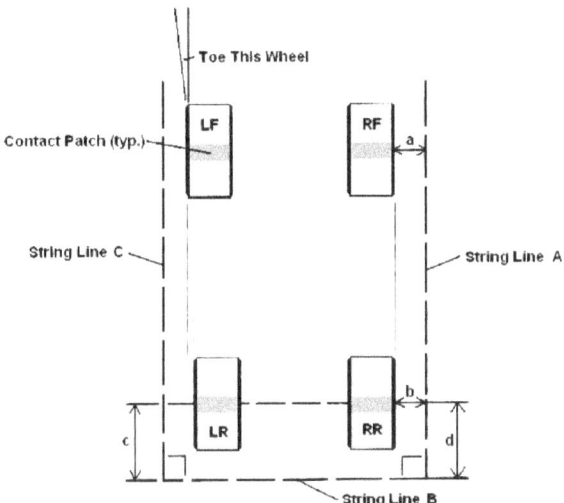

*This is what we should end up with after aligning the rear to the right side tire contract patches for a circle track race car. Measurements "a" and "b" should be equal at the contract patch, as well as dimensions "c" and "d". We don't need to be concerned with the alignment of the left side tires for most circle track applications. There is much less loading on those tires through the turns and less influence on the handling of the car due to miss-alignment. For road racing cars with solid axle rear suspensions and formula type cars, we place the tire contact patches equidistant from the centerline of the car, or from the offset lines running outside the car.*

### Camber Offset Chart

| Deg. of Camber | Tire Circumference | | | | | | |
|---|---|---|---|---|---|---|---|
| | 80" | 82" | 84" | 85" | 86" | 87" | 88" |
| 2.0 | 0.444 | 0.455 | 0.466 | 0.472 | 0.478 | 0.483 | 0.489 |
| 2.5 | 0.555 | 0.569 | 0.583 | 0.590 | 0.597 | 0.604 | 0.611 |
| 3.0 | 0.666 | 0.683 | 0.700 | 0.708 | 0.716 | 0.725 | 0.733 |
| 3.5 | 0.777 | 0.797 | 0.816 | 0.826 | 0.836 | 0.845 | 0.855 |
| 4.0 | 0.888 | 0.910 | 0.933 | 0.944 | 0.955 | 0.966 | 0.977 |
| 4.5 | 0.999 | 1.024 | 1.048 | 1.061 | 1.074 | 1.086 | 1.099 |
| 5.0 | 1.110 | 1.137 | 1.165 | 1.179 | 1.193 | 1.207 | 1.221 |

*The tire contact patch at the ground is offset from where the contact patch is located at the hub height. We need to know what this offset amount is when the wheel has either positive or negative camber. The chart shows how far the contact patch is offset from vertical due to wheel camber from hub height. If the front camber is in the negative direction (the top is closer to the centerline of the chassis), we subtract that offset distance from any measurement we make to the rear axle tire sidewall at hub height to align the rear to the front. For example, if we measure 10 inches to the front sidewall at hub height and the negative camber offset is 0.5", then when we measure to the rear tire sidewall (assuming it has zero camber) at hub height, we subtract the 0.5" from 10" to measure 9.5" to the rear tire sidewall so that the front and rear tire contact patches at the ground will then line up. If we just measured 10" to each front and rear sidewall, then the tire contact patches would not line up because of the front wheel camber.*

To align the rear tires to the front tires for solid axle rear suspensions, adjust the panhard, J-bar, or Watts links, length so that the right side tire contract patches are: 1) the outside (the turns) contact patches are in-line for circle track cars, or 2) each of the two rear contact patches are equidistant from the centerline of the chassis for road racing cars.

NOTE: The rear wheels may also have camber set into them. Look at the chart to estimate how much to compensate for the cambers. Find the difference in the compensation amount and the offset read at the wheels to find how far from the string the wheels need to be in order to line up the tire contact patches. Negative camber settings will require subtraction of the camber offset from the hub height offset. Positive camber settings will require adding the camber offset amount to the hub height offset.

As we adjust the rear end side to side in a solid axle rear suspension, there is a possibility that the rear end may not be square to the centerline and will change as the rear end moves laterally. That is why we do the rear wheel front to rear alignment first. Now that we have aligned the contact patches the way we want, we will now need to check to make sure the rear end is perpendicular to the chassis centerline for the solid axle suspension and if not, adjust it.

**Step 5 - Rear End Squaring Alignment** – Once the front to rear tire contact patches alignment has been completed, we can then square the rear end. The rear wheels should always be set parallel to the centerline of the car for most types of racing. For road racing and formula type cars, there is a desired amount of toe-in usually designed into the alignment of the rear wheels. This toe-in can now be set.

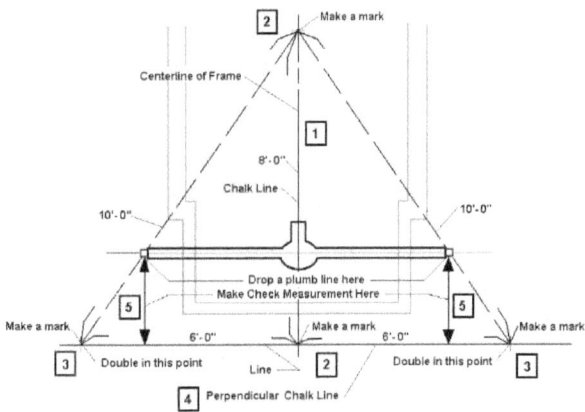

*Here we see the layout for the process of setting up to square the rear end. Note that the lines formed by the intersecting marks are right triangles. We use 6-8-10 foot sided triangles. Take your time and measure very accurately. Re-measure if you feel that everything is not right once you pull the rear string line across the outer points of the triangles. A line over points 3 should align with point 2 on the rear perpendicular line.*

For solid axle rear suspensions, we simply set the rear axle to be perpendicular, or at ninety degrees, to the centerline of the car. We do this by creating a line that is perpendicular to the centerline we have already established. Using a simple 3-4-5 right triangle with the lengths double to 6-8-10 feet, we can measure using the centerline to establish the perpendicular line that will be used to measure to the rear end. We simply make a mark on the centerline under the car and measure 8 feet back down the centerline to the rear and then make a second mark, the rear point being placed well behind the rear end.

We then set an outer point by measuring from the rear point 6 feet out and 10 feet from the front centerline point to establish a point on each side. A line between these two outside points forms a line that is ninety degrees off of the centerline that we can then measure from to the rear hubs to square the rear end.

Now all we do is drop a plumb bob line down off the front or back of the hub on each side and measure to the rear line. Those two measurements must be the same. If not, adjust the trailing arm lengths so that they are equal. Now you have squared the rear end and it is aligned properly.

**Summary** - Now we should have all of our wheels pointed straight ahead and parallel to the centerline of the chassis. And, our contact patches should be aligned so that either the front and rear tire contact patches are equidistant from the centerline for road racing and formula cars, or for solid axle circle track cars, with the outside contact patches aligned, or offset if so desired.

When we know that the wheels are pointed where we intend for them to be and the tire contact patches are the desired distance off the centerline, we can then know that we can proceed with the other setup settings that will not be affected by miss-alignment of the wheels.

# Exam - In The Context Of This Lesson:

### If Our Alignment Is Wrong, We Can Fix It How?

1) By using different spring rates

2) Through shock developement

3) By correcting the alignment issue

4) By changing the weight distribution

### For Circle Track Racing, Where Do The Rear Wheels Point?

1) Into the turns

2) Straight ahead on entry and through the mid-turn

3) To the right of centerline

4) To the left of centerline

### The Proper Tool To Use To Align The Race Car Is?

1) The string, it never lies

2) A hub mounted laser system

3) A remote mounted laser system

4) All of the above

### When Using The String Method, What Is The First Thing We Do?

1) Find the high point or runout in the tires

2) Center the steering

3) Point the front wheels straight ahead

4) Square the rear end

### When Doing Alignment, We Set The Front Toe To?

1) An eighth inch out

2) An eighth inch in

3) To zero toe

4) What works for our race track

### Rear Solid Axle Alignment Is Always?

1) Perpendicular to centerline

2) Lining up the right side contact patches for circle track

3) Centering on the centerline for road racing

4) All of the above

# Racecar Technology – Level Two
## Lesson Nine – Alignment Rear with Roll Steer

Now that we have the wheels aligned and the tire contact patches where we want them, we need to address the alignment of the wheels as the chassis dives and rolls through the turns. There may well be a different alignment once the chassis moves. For cars with AA-arm suspensions, front only or front and rear, the steer for that system would result strictly from bump steer, which we already talked about in this Course.

The steer and alignment that we want to concentrate on in this Lesson is what results from chassis motion that commonly exists in the straight axle suspension systems. That is what this discussion is all about. Most of these types of suspensions, with their various attachment designs, will steer as the chassis dives and rolls through the turns.

*Rear steer is mostly desired in dirt track racing. Here we see two different angles for each of these cars. The 45 car has much more rear steer (notice the LR tire moved forward in the wheel well) than the 0 car. The rear steer is caused by the angle of the links that hold the rear end in the car and the movement of the chassis pushing and pulling on the links to in-turn push and pull the rear end. Note also that the greater rear steer causes a larger angle of the chassis to the direction the car is traveling. The 45 car is "hanging it out" much more than the 0 car. There is a reason why dirt teams do this.*

It is very important to know how this happens and how we can dictate when and how much our rear end steers as we enter and drive through the turns. The presence and amount of rear steer needed is very dependent on what type of race car it is.

For asphalt race cars competing on circle tracks or even road racing, we neither need nor want very much rear steer at all. The rear of these straight axle cars are very sensitive to the direction the rear tires are pointed. Even if we desire and incorporate rear steer, a little goes a long way.

For dirt cars, it has been the trend to cause a lot of rear steer for various reasons. We'll get into those cars designs and see where we can cause steer that might be beneficial to performance. Let's now discuss how rear steer happens and how we can design our cars to steer, or not.

**What Is Rear Steer?** - Rear steer in a race car is a geometric effect that is mostly caused by suspension movement. Under the right conditions in measured amounts, RS can be beneficial and enhance performance. Under the wrong conditions, it can ruin your handling. We need to have a solid understanding of what produces RS and what effect RS has on the handling in our cars in order to have the best performance.

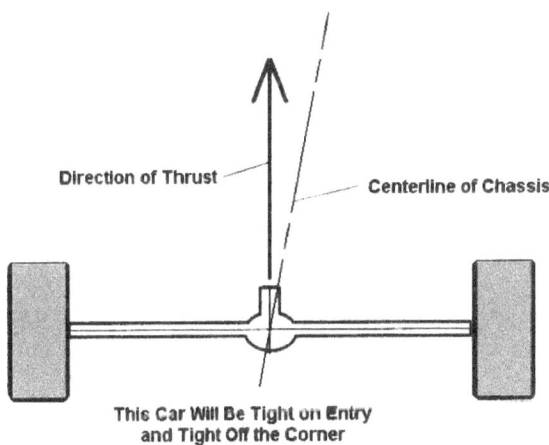

*With the rear end pointed to the left of centerline, the steering will cause the rear of the car to want to run under the front end causing a very tight condition, especially under acceleration. With the rear end pointed to the right of centerline, the car will be freed up going in, through the middle and possibly loose off the corners with the rear end wanting to run around to the right of the front.*

When we have rear suspension movement, as the rear corners of the car move and the control arms that locate the rear end fore and aft move, the components for each side can cause the wheels to either move forward or to the rear.

A rear end that is steered to be pointed to the left of centerline will cause the thrust angle to be left of centerline and that makes the car tighter on entry and tighter on exit under acceleration. A rear end that is steered to the right of centerline will cause the thrust angle to be pointed to the right of centerline and that makes the car loose on entry and loose on exit under acceleration.

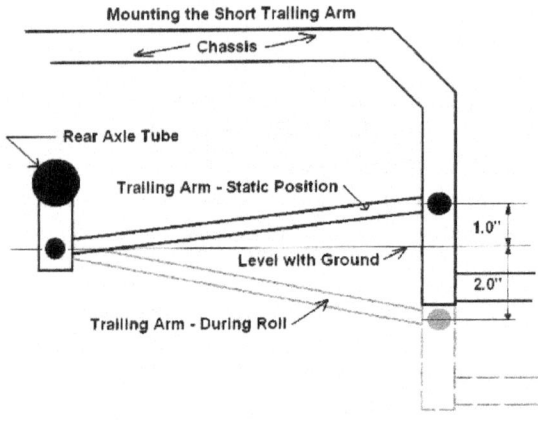

*The three link rear suspension system can produce rear steer in both directions. As an example, let's say the right rear of the chassis moved down 3.0" through the turns and the front mounting point was 1.0" higher than the rear mounting point. Then as the chassis moves down on the right side 1.0", the RR wheel will be moved back as the front mount approaches the height of the rear mounting point. As the front mounting point continues to move down, the RR wheel will be pulled forward and return to the original fore and aft position as the front mount goes 1.0" down from level. Normally, to produce a small amount of rear steer to the left on asphalt, we mount the front pivot point 1/3 of the total travel distance higher than the rear pivot. In this case, the chassis traveling another 1.0" to make up the total of 3.0" of travel continues to pull the rear end forward a slight amount to create the rear steer to the left.*

The asphalt racing surface provides a lot of traction, even on those flat "slick" tracks. Because there is very little slip of the tires on asphalt, the range of useable rear steer is very small. There are six predominant rear suspension systems used in stock car racing and all of them produce some amount of rear steer. They are:

**The Three Link System**– This system has two trailing arms mounted near the rear tires under the axle tubes and one third link, usually mounted on top of the rear differential. The top, or third, link controls rear end wrap-up. The trailing arms can be permanently mounted from a top view to be parallel to the centerline of the car or angled with the front mounts closer to centerline. They can also be designed to be adjustable to create different angles from a side view.

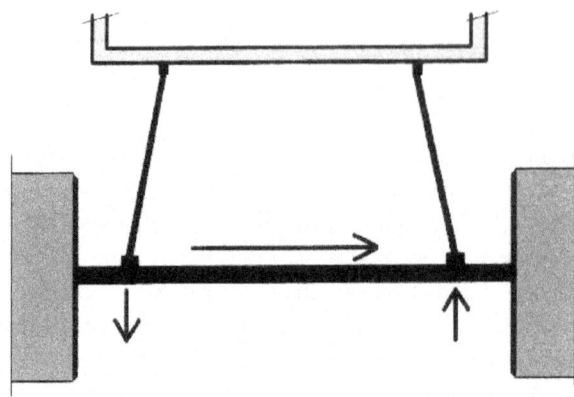

*Here we see the lower links in a three link suspension. The front mounts are attached to the chassis to be closer to the centerline of the chassis than the rear mounts. With this design, if the rear end were to move laterally, there would be rear steer. If the rear end moved to the right, the side would move forward and the left side would move rearward. This causes rear steer to the left which would tighten a car turning left. If the rear end moved left, the opposite would occur, and the effect might loosen the car. This is important to think about with a panhard bar system. Which way does the bar push the rear end as the chassis travels?*

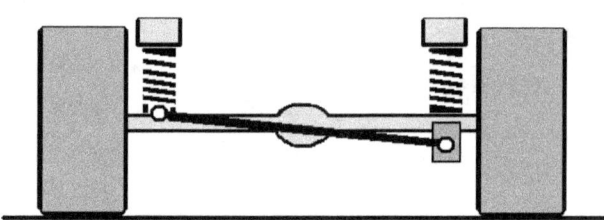

*It is a result of panhard bar angle, and the motion of the chassis in roll, that can either pull the rear end to the right or push it to the left. If the panhard bar were mounted with the right end (chassis mount) lower the left side that is mounted to the rear end axle housing, then with chassis movement down resulting from chassis roll, the rear end would be pulled to the right. If the bottom two links in a three link system were like the top view angles in the previous illustration shows, then the rear end would move into steer to the left. If on the other hand, the panhard bar were mounted with the right end higher than the left end in this same car, then the motion of the chassis would push the rear end to the left and that would produce rear steer to the right. This would act to loosen the car, but also cause an angle to where aero flat plate effect might just play into causing a force to the inside of the turn helping to go faster through mid-turn. Many of the modern asphalt late model circle track cars have mostly flat sides that would help create this effect.*

RS in this system is caused by chassis movement which can produce several secondary effects. Usually, the right rear corner of the chassis moves more than the left side, and on most flat to medium banked tracks, the left rear moves very little.

On most asphalt 3-link cars, the RR trailing arm mostly controls RS due to body roll. We usually need to position the angle of the trailing arm so that the front mount is higher than the rear mount by roughly one-half to one-third the distance that the front mount will move down during cornering. This compensates for the movement of the front end of the link and in reality pushes that side of the rear end back first and then pulls it back to where it was resulting in very little rear steer.

**The Truck Arm System** – The truck arm system has been adapted from the design for a 1964 Chevy pickup trucks and is used on many Late Model Stock cars as well as the three premier divisions of Nascar, being the Camping World Trucks, the Xfinity cars and the Cup cars. These systems only steer to the left and have a limited amount of steer. The roll of the chassis and the movement of the panhard bar are the two components that influence the amount of steer in these systems.

As far as geometry related to rear steer is concerned, this is a good system for asphalt cars. The amount of rear steer due to body roll is regulated by the height of the front mounts of the arms which are always mounted lower than the rear point of rotation which is the axle. Rear steer amounts due to the panhard bar vertical movement are regulated by the angle of the bar.

**The Standard Dirt 4-Bar System** – The 4-bar suspension is highly adjustable and can be made to steer in either direction. The rule about never steering the rear end to the right on an asphalt car does not apply to a dirt car. There are times when we definitely want the rear to steer to the right.

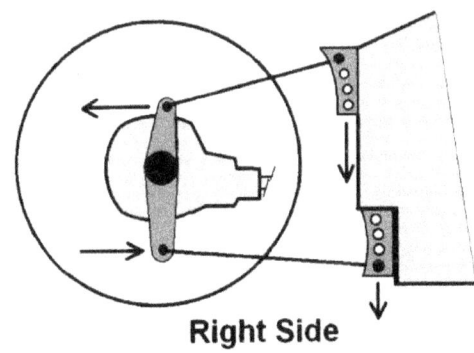
**Right Side**

*If the links on a 4-bar car are set in certain predetermined holes, the movement of the top and bottom mounts at the rear end will compensate for fore and aft movement that results in zero rear steer. Most of the existing rear steer comes from movement at the left side of the chassis and the mounts located there.*

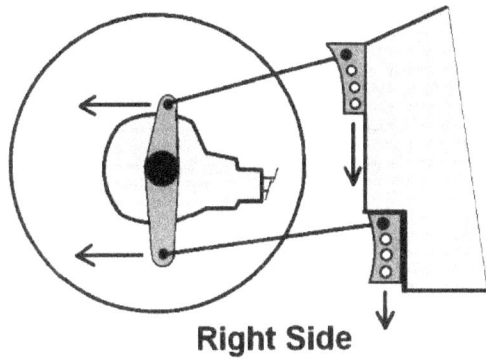
**Right Side**

*Here we can see how, on the right side of a four-link system, mounting the forward ends of the links in the top holes causes movement of the right side of the rear end back when the chassis moves down. On the left side, the opposite happens when that side of the chassis moves up and the rear end moves forward.*

To determine the amount and direction of rear steer in this system, we must know the movement of the corners of the car. On dirt cars, this can be very different than what we see on asphalt cars.

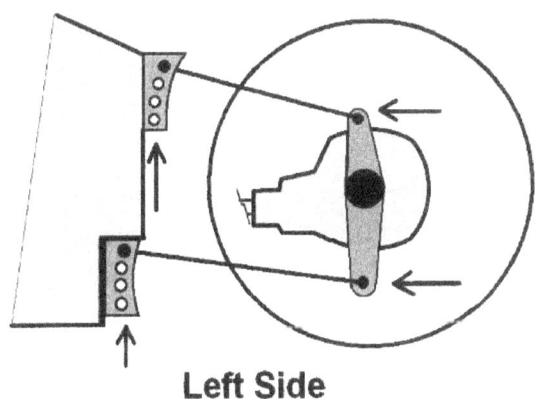
**Left Side**

*The way a four-bar car steers so much is by the use of high angles in the links on the left side of the chassis. When these links forward ends are mounted higher than the axle mounts, the motion of the chassis rolling up going through the turns plus the force of the thrust of acceleration pushing the chassis up cause the left rear wheel to move forward to create a lot of rear steer. This pushed the rear of the car out to the right by necessity in order for the rear wheels to line up with the direction of travel. The flat plate aero created by this angle of the body helps push the car to the inside of the turns and helps it go faster, especially on dry slick tracks that have low grip.*

**The Metric 4-Link System** – The metric four link is a widely used system that comes on some models of stock automobiles. It uses four links as the name implies that are not parallel to the centerline of the car. The top links are angled from a top view with the front pivots wider than the rear pivots. The lower links are angled from a top view with the front pivots narrower than the rear pivots.

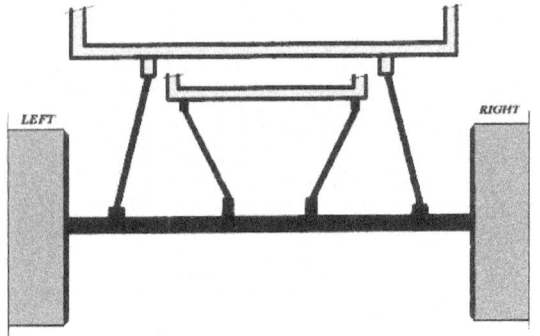

*The Metric 4-link suspension used on many models of street stock and some trucks has two links above the rear end and two links below the rear end. They are angled from a top view to prevent the rear end from moving side to side as the chassis experiences lateral loading through the turns. These rear suspensions have a traditionally high rear moment center. The rear steer is dependent on the ride heights in the rear which determines the link angles.*

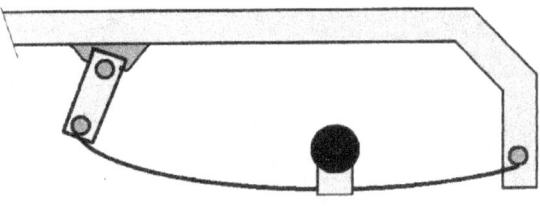

*We can see how the front and rear chassis mounts in a leaf spring suspension act somewhat like the links in a three link system. The rear mount swings in a shackle and has no effect on rear steer, or movement of the rear end. But the front mount can move the rear end on that side fore or aft depending on if it is higher or lower than the center of the axle. If higher, it would push the axle back as the chassis moves down. If lower, it would pull the axle forward as the chassis moves down.*

With this system, the rear end stays located by virtue of the opposing angles of the upper and lower links. There may be rear steer inherent in this system depending on the ride heights because the original car was designed for stock ride heights. When we race a stock car, we usually change the ride heights. Under most current rules, there is no adjustment for amounts of rear steer with these systems.

**Leaf Springs Systems** – The leaf spring rear suspension system locates the rear end fore and aft as well as laterally using only the leaf springs. There can be a small amount of rear steer as the chassis rolls and squats, but it is both minimal and mostly fixed as far as adjustability. The advantage of this system is that it keeps the rear end squared up and the thrust under acceleration more straight ahead, if that is what is needed.

The height of the front eye of the leaf spring in relation to the axle centerline determines the amount of rear steer. Spacing the leaf different heights from the axle tube can be a way to change the height relationship of the front eye and the axle, which will change the amount of rear steer for this system.

**The Z-link System** – The Z-link rear suspension, or swing arm as it is sometimes called, is another system used on dirt cars. Compared to the 4-bar cars, it has more limited adjustment for rear steer and historically has worked well on the tighter and more highly banked race tracks because the direction the rear end is pointed is more straight ahead.

Some manufacturers have added multiple mounting points on the front and rear chassis mounts. This helps make the rear steer characteristics more adjustable for the changing conditions.

*A Z-link suspension system uses a link extending from the under the rear end forward to the chassis and one from the top of the rear end rearward to a mount on the chassis forming a Z pattern. Most designs use many mounting holes that enable the team to adjust for the amount of rear steer in those cars. You can set your angles and then move the chassis to determine how much the axle is pulled forward, or pushed backwards. This helps to determine how much rear steer is in a certain system. There is no right or wrong in these designs, but experimentation at different race tracks helps teams find the right combination that will work for both the track conditions as well as what the driver feels comfortable with.*

**Measuring Your Rear Steer** – To measure the rear steer in your car, first know the shock travels at each corner of the rear. Then you can recreate the rear attitude at mid-turn. You can use a tape measure for simple measurements or a laser system for more detailed analysis. With the tape, simply measure, at ride height, from the rear wheels towards the front of the car level with the ground to a point. This can be a mark on a tape near the front wheel. For the laser systems, follow the manufacturer's instructions.

Move each of the rear wheels in any increment you choose ending up in the position they are at when the car is at mid-turn. and measure the fore and aft distances at each increment. If the movement on the right and left are in opposite directions, add them to determine the amount of rear steer. If they are in the same direction, subtract them. You might be surprised at how much rear steer is going on.

**Why Do We Want Rear Steer?** – Dirt cars need all of the aero help they can get. There is an aero effect called Flat Plate aero. That means, if a flat plate were traveling through air at an angle, it produces a force. If the front of the plate were lower than the rear of the plate, then that force would be directed down. The air being pushed out of the way exerts a force on the plate in that direction the plate is pointed in.

If the flat plate were mounted, or positioned vertically, then the force would be exerted laterally. If the front of the flat plate were positioned left of the rear, then the force would be directed to the left. A cars body can be made to create an aero force in a lateral direction.

A modern late model dirt cars body is flat on the left side. That side acts as a flat plate and when driven through the air at an angle with the front left of the rear, it exerts a force to the left. If that car turns left, then the aero force helps it resist lateral forces that try to run it off the track in the direction of the lateral force.

So, why do we need rear steer in a dirt late model? It's not related to helping the car turn in or help it point off the corners like some people think, it is to help provide the lateral aero force pushing to the inside of the turn needed to resist the lateral force trying to push the car to the outside of the turn.

Remember too that there is a limit to how much angle you can use to create the aero sideforce. Too much angle and the effect is diminished and the sideways attitude creates a lot of drag slowing the car down.

On asphalt, we can utilize rear steer in smaller amounts than what is used on dirt for a similar reason. If the rear were hung out a bit, there would also be a lateral aero force helping to keep the car on the track. Everything helps and if that force can help us go a little faster, even better.

But along with that steer to the outside, we need to be able to get off the corner and steer to the outside of the turn is detrimental to getting up off the turns under power. In the next Lesson, we'll tell you how to have your rear steer and get off the corner with rear steer under power.

**Summary** – Rear steer design is something we need to know about and utilize, but telling you exactly how much is like telling you exactly how far to go into the corner before braking. Every track is different and every car is somewhat different, even in the same class.

So, in this Lesson, we wanted to provide enough information about how to create rear steer and why so that you can make an intelligent decision as to how much would be good for you. The rear steer we get from chassis dive and roll is not the entire picture of rear steer. It is only the beginning. In our next Lesson we will describe another type of rear steer that can be utilized along with the dive and roll rear steer.

# Exam - In The Context Of This Lesson:

### Rear Steer Is Defined As The?

1) Rear direction of the tires when we turn

2) Change in direction of the rear tires when the chassis moves

3) Alignment at static ride height

4) Alignment at dynamic ride height

### For Asphalt Race Cars, We Need?

1) A lot of rear steer to make the car tight

2) Rear steer to the outside of the turns

3) Little or no steer at all

4) The rear wheels to steer with the front wheels

### For Dirt Race Cars, We Need?

1) Rear steer to the outside of the turns

2) More steer for slick tracks

3) Less steer for tight tracks

4) All of the above

### In A Three Or Four Link Rear Suspension With Straight Axle, The Rear Will?

1) Never move through dive and roll

2) Will always steer through dive and roll

3) Will always steer to the right

4) Will always steer to the left

### For Dirt Race Cars, We Use A Lot Of Steer Why?

1) To make the car tighter

2) To make the car looser

3) To make the car run at an angle for better aero

4) To help turn the car

### Truck Arm Systems Always Steer?

1) To the left

2) To the right

3) Very little at all

4) Quite a bit

### Steer In A Leaf Spring System Is Determined By?

1) Height of the rear spring eye

2) Height of the front eye

3) Angle of center of axle to the center of the front eye

4) Angle of rear eye to front eye

### We Use What To Measure Rear Steer?

1) The shock travels

2) Chassis roll angle

3) Chassis dive

4) All of the above

## Racecar Technology – Level Two
## Lesson Ten – Rear Steer Under Power

In the last Lesson we learned how to utilize rear steer that comes from chassis dive and roll. We talked about the straight axle suspension system exclusively because that is the only system we can use to create significant rear steer. The AA-arm suspension systems don't work well to produce rear steer, and we wouldn't want them to. So again, this discussion involves only the straight axle rear suspension related to rear steer, and now as it is created while under power.

*Some classes of dirt cars use quite a bit of rear steer to provide some advantages for getting around the corners. The car is not loose, as in the rear tires are sliding, they are steered this way on purpose.*

To refresh, rear steer can occur while the chassis dives and rolls through the turns. We can use this steer in various ways depending on the type of race car and the needs of the team. The other form of rear steer happens while we are applying power when accelerating off the corners. We use the torque of the engine and the thrust of acceleration to create rear steer independently of the rear steer we get from dive and roll.

In fact, the RS we get from acceleration can counteract the rear steer we got from dive and roll. That is an interesting concept and one we will explore further. In my opinion, very few race teams know how to properly utilize RS. Of all chassis related performance areas, Power RS is second in importance being just below mid-turn dynamic balance.

We use mid-turn dynamic balance to make sure the loading on the tires is correct and what the car wants, and we use RS from dive and roll to help the car stay on the track. But at some point we need to accelerate off the corners and the RS we have to get us through the middle of the corner can be detrimental to having good bite off the corner. In other words, we need the opposite RS for corner exit.

**How Does Power Induced Rear Steer Happen?** – Three things are happening when we accelerate that can be used to create RS under power. One, we have weight transfer from the front to the rear. As this weight transfers, it puts more load on the rear springs and they can be made to compress causing the rear of the race car to move downward. This front to rear weight transfer only happens while under acceleration and therefore doesn't affect the mid-turn, steady state handling.

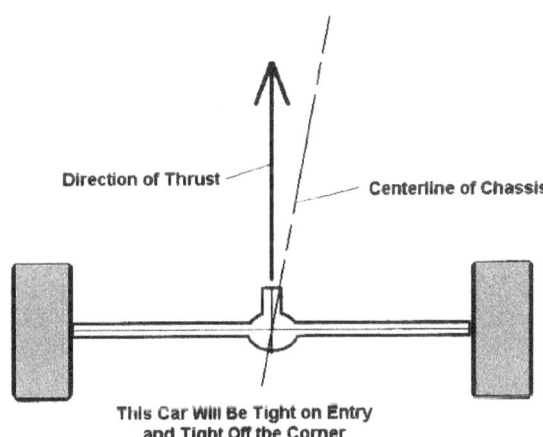

*A race car steered this way, with the rear tires pointed to the left of centerline, will always be tight. The front must try to steer to compensate for the rear of the car moving left. This is almost never desirable.*

The second thing that happens is that we have the torque of the motor providing thrust to push the car faster, and so there are uses for that thrust to create rear steer only under acceleration. And the third thing that occurs while under acceleration is that engine torque is also trying to twist or rotate the rear end as the pinion tries to climb the ring gear in the rear end. The rotation is clockwise when viewed looking at the left side tire and we can use the rotation to create rear steer again only while under acceleration off the corners.

OK, all of that sounds pretty cool because what we do to the setup and design of the race car for better performance off the corner must not upset or interfere with the mid-turn handling. Remember, our number one goal is to create the ideal mid-turn dynamic balanced setup. Anything we do to the chassis to improve, or solve problems with exit performance that changes the mid-turn performance takes us backwards in overall performance.

Now that we have three components we can use to enhance corner exit performance, without interfering with the mid-turn performance, let's study each one, one at a time. These effects can be mixed or used individually. As with any performance adding technology, more is not necessarily better. There are limits to how far we can go with our use of RS.

**Weight Transfer Utilization** – The use of weight transfer to provide better bite off the corners has long been in use. Older circle track setups often used softer right rear (outer rear springs) springs than the left rear (inside) spring. When the weight transferred and the rear of the car compressed, the left rear with the stiffer spring compressed less and gained more of the weight that had transferred.

This increased the cross weight percent and tightened the car when exiting off the corners. As setups became more modern and teams started running stiffer RR (outside) springs, the effect was diminished or reversed, so that option became un-useable.

RS to the left (inside) which definitely will create more grip in the rear tires.

Along with that effect, installing a shock in the LR corner of the chassis that is stiffer in compression settings than the one in the RR corner will also load the LR tire more, but only while the shock was moving as the weight was transferred onto the spring. Once the shock stopped moving, the effect goes away whereas the RS stays as long as the car is accelerating.

This RS effect is significant because on most flat and medium banked race tracks, with modern circle track cars and even with road racing cars, the inside rear corner does not move vertically much from where it is statically at ride height. Therefore, any angle we put in the link to the corner has no affect until that corner moves and that movement is mostly while under acceleration.

**Thrust Created Rear Steer** – As we talked about above, when the car accelerates, there is thrust created that causes the rear end to push against the links and the chassis. This thrust is significant. To accelerate a 2800 pound car from 60+/- MPH up to 120 MPH takes a lot of thrust. If we measured it in pounds of thrust, it would come in at around 1,500 to 2,000 pounds total. If we divide that between the two rear tires like we should, each tire would be pushing at 750 to 1,000 pounds. That is a lot.

*When we add angle to the left trailing arm in a three link suspension system, when the car squats under acceleration, the angle will drive the left rear back causing rear steer to the left for a left turning race car. We can vary the amount of rear steer by changing the trailing arm link angle.*

Later on, teams started utilizing the rear link angles to help create RS. As the car squats from the result of added loading from the weight transfer, adding angle to the left rear (inside) link, with the front mount higher than the rear mount, on a three link suspension caused the left (inside) wheel to be pushed back. This created

*One way to create RS is to install compressible links in a three or four link suspension so that when the car accelerates and creates thrust, the link will compress and that side of the rear end will move forward creating RS. This rear steer almost always points the rear end to the inside of the turn to create more angle of attack for the rear tires. This creates more grip, or side force enabling the driver to use more throttle without the rear tires slipping out.*

This concept can be used in reverse as well. With some stock classes, a compressible link or mount can be made to work both ways, compressing while under acceleration and extending while under braking. Both produce RS, the first pointing the rear end to the inside of the turn, the second pointing the rear end to the outside of the turn helping the car to turn.

I've personally seen this used in a stock class at a tight, flat turn race track. That car was dominant and no other car was doing anything like that. For those of you who are racing in stock classes, or know someone who is, this little gem might just give you a huge edge if done right.

*The three link rear suspension, and the lift arm and torque arm suspensions all create AS. On the three link, the link is pulled under acceleration, and the lift and torque links forward ends are lifted. When we allow the rear end to rotate a small number of degrees, we can use that rotation to create RS.*

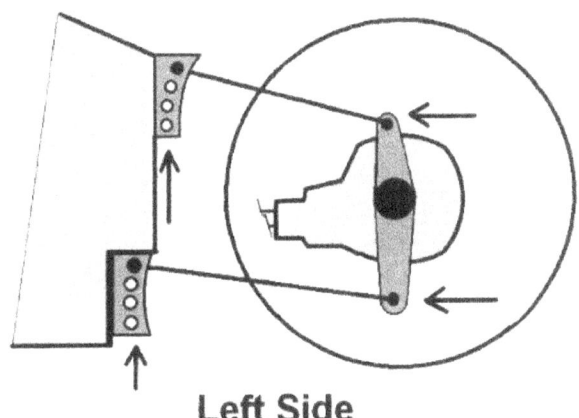

*When the left side links on a four-link suspension are in the top holes, as the car accelerates and the rear wheel drive forward, the angle of the links pushes the chassis up at the left rear and forces the right front corner of the car down. This is a desirable attitude for maximizing the aero efficiency of the car.*

**Torque Induced Rear Steer** – The other effect that can be used to create RS only under acceleration is from the torque of the motor trying to rotate the rear end. As the pinion gear in the rear end pushes against the ring gear, it tries to climb the ring gear and as a result pushes the rear end housing and tries to rotate it. This rotation is clockwise when looking at the left rear tire.

We do that by installing a compressible spring or rubber type of device in the third link or lift arm that would compress and allow the rear end to rotate. When it does rotate, steer is created by staggering the height of the side links and the distance from those mounts to the center of the axle.

When the rear end moves in rotation, the axle moves back because it pivots on the bottom link mounts, but if the left (inside) link mount was lower than the right (outside) link mount, then the left side of the axle would move farther back creating rear steer to the left, or towards the inside of the turn.

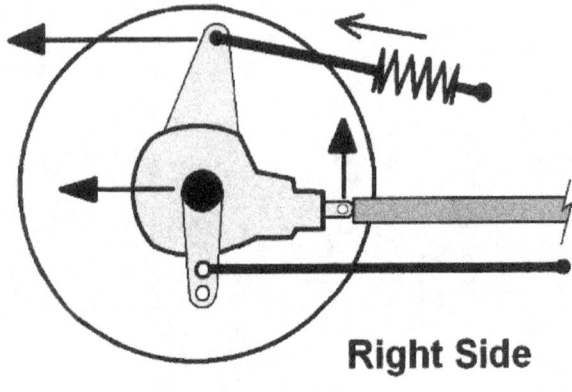

**Right Side**

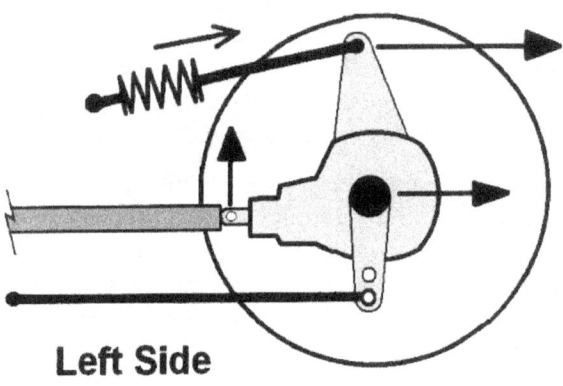

**Left Side**

*This type of rear steer is utilized mostly on asphalt race cars and only with circle track cars because it only works when accelerating off of a turn in one direction. It would work the opposite way, or the rear steer would be the other way, if accelerating off a turn to the right making the car have less grip rather than more.*

The actual amount of steering, or degrees of rear end change in direction, is very small, but then again, asphalt race cars are very sensitive to any amount of rear steer in any suspension system. In practice, we usually see where the left rear tire will move back 0.050" to 0.075". The angle created is about one tenth of a degree. To put the numbers into real tape measure dimensions, the left rear tire is moving a little less than, or more than, one sixteenth of an inch.

**Roll Steer Plus Power Rear Steer** – We talked about RS created by chassis dive and roll in the last Lesson, and now we can talk about using both RS types to enhance mid-turn and exit performance. Let's start with asphalt cars because they are a little different in needs from dirt cars.

On asphalt, we might need a slight amount of rear steer to the right (inside of the turn) to allow the car to rotate into and around the corner and lessen the amount the driver has to turn the steering wheel. This is, or could be, a two edged sword so to speak.

If the RS to the right did indeed help us into and through the middle of the turn, then surely it would make us loose off the corner because the thrust angle would be right of centerline and that would want to take the rear of the car towards the outside of the turn. We would quickly learn this is not a good thing if we didn't lose the car in the process.

What if we could loosen the car into the corner and through the middle, but tighten it off the corner by using RS to the left (inside) to counter or reverse the initial RS to the right? From what we learned in the bulk of this Lesson, yes we can do that. And winning teams are doing that right now, we just don't hear much about it, until now.

For dirt cars, especially the late models and modified classes, rear steer is being used to create aero side forces, as we have explained, to enhance the lateral force working against centrifugal forces trying to push us off the track in the turns. That is all well and good, but when we accelerate off the corners, we don't need as much aero help anymore since we are going more straight and not turning.

It is at this point that power induced RS can be utilized to bring the rear end back to pointing more inline with the centerline of the car. This too is a two edged sword because with those cars, the driver is sometimes on the throttle through mid-turn and the RS would be working back and forth and possibly destabilize the car.

Only the team can determine if and when RS either way would work to enhance performance. Our job in this school is to teach you methods and give you ideas, not setup your car. There are too many variables involved to say it has to be one way or another.

# Exam - In The Context Of This Lesson:

### Rear Steer Under Power Can?

1) Add to the mid-turn rear steer

2) Subtract from mid-turn rear steer

3) Be designed to be independent of mid-turn rear steer

4) All of the above

### One Way To Create Power Rear Steer Is?

1) Using the squat of the rear from weight transfer

2) Using the acceleration force driving the car forward

3) Using the torque of the motor

4) All of the above

### In A Three Link Suspension, More LR Angle Will?

1) Cause rear steer to the right under acceleration

2) Cause rear steer to the left under acceleration

3) Help keep the rear end straight

4) Make the car loose off the corners

### In A Three Link Suspension, More Third Link Angle Will?

1) Cause more squat and rear steer

2) Cause less squat and rear steer

3) Reduce weight transfer

4) Increase weight transfer

### In A Four Link Dirt Suspension, More Left Side Link Angles Will?

1) Cause more rear steer under power

2) Help produce aero side force

3) Load the LR tire under acceleration

4) All of the above

### Power Rear Steer Should?

1) Be used to correct corner exit problems

2) Always add to mid-turn performance

3) Take away from mid-turn performance

4) Be used to correct mid-turn handling

## Lesson Eleven – Toe Front and Rear

This Lesson is a short and concise review of what Toe is in a suspension system and some theory on why we have it and how to set Toe. Basically, Toe is a feature of a AA-arm suspension that adds stability to the car. If is defined as: the miss-alignment of two tires/wheels on a suspension system. Perfect alignment is what we know as both wheels being pointed in the exact same direction. With Toe, we cause them to point in slightly different directions.

To further define Toe, Toe-In is when the front of the tires are closer together than the rear of the tires. Toe-Out is the opposite. Normal front Toe-In for most applications is from 1/16" (0.063") to 3/16" (0.188") of Toe. Some teams have actually used ¼" (0.250") of toe for very short ¼ mile tracks, but based on calculations, that is too much in our opinion.

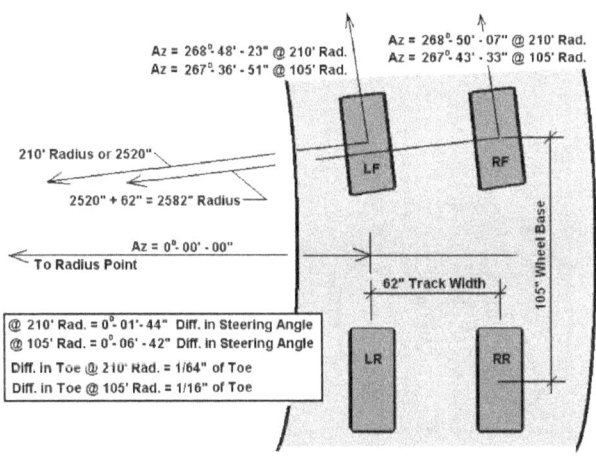

*These calculations show that we need very little Toe-Out in the front wheels in order for each wheel to track correctly along its individual arc. The different arc radii create the need for small amounts of Toe.*

We need Toe in our race cars to stabilize a AA-arm suspension, be it front or rear. In a straight axle suspension, we usually do not use any toe. The amount of toe is similar to the amount of Ackermann we learned about. We need very little. For most double A-arm applications, we need from 1/16 inch to ¼ inch of Toe-out for most circle tracks and road courses. For the rear AA-arm suspension, we usually need Toe-in, say from 0.020 to 0.035 inch of toe depending on the race track.

**How Much Toe Do I Need** - The general rule is that you need less front Toe-Out for: 1) larger radius turns, 2) Higher banked tracks, 3) very fast tracks with large radius turns. As for rear Toe-In For a AA-arm suspension, much smaller amounts are used. These range from 0.020" to 0.035" in most cases.

It has been learned through the years that this miss-alignment does stablilize the car. And the amount and direction of the Toe is dependent on the type of suspension. A race car with rear wheel drive and a AA-arm front suspension usually uses Toe-Out in the front. For a rear AA-arm suspension such as on a prototype or formula car, the Toe is usually set to Toe-In.

*Like we said, for a straight axle rear suspension, we need very little, or no, toe. This rear end has a device that can be used to straighten the rear end and/or help it to stay straight under acceleration by use of braces coming off the center section.*

Proper toe settings at the front and the rear are very important. A set of tires that are not toed correctly will create a lot of drag, much like applying the brakes and sometimes more efficiently than the actual brakes. The standard of the circle track industry is to toe the front out 1/16" to 3/16". The rear wheels are to be straight ahead and parallel to each other, having no toe at all.

*We can use lasers to measure toe by placing a target at a distance in front of the wheel hub and another one to the rear at the same distance. If we put the targets far enough apart, then the toe will be magnified and make it easier to set exact toe amounts. We commonly use toe plates, but we must be careful to read the tape measure accurately.*

Some theory exists that the tire contact patch deflection dictates the toeing of one or more of the rear wheels to compensate for this, but it is hard to justify and can probably be best defined as a crutch used to help solve some other setup miscue.

We can set the toe using one of three different methods. We can use simple toe plates that rest against the tire sidewalls. These have two slots, one at the front and one at the rear about an inch off the ground. We run a tape measure across the rear between the slots and remember that number. Then we measure at the front between the slots and subtract to find the Toe. There can be errors in this method from tire/wheel runout, sidewall protrusions like lettering, etc. and sloppy handling of the toe plates. This procedure should be done several times until the toe reading is consistent.

We can also use strings, but we need to makes sure the strings are parallel on each side of the car. And, the strings must be small diameter, preferably heavy fishing line or similar. We have used this method on prototype cars for setting the rear toe.

Finally, we can use lasers. This method could be the most accurate because we can place the laser far to the front of the wheels where the Toe amount is multiplied. If we set the targets four times the diameter of the tire, then we can easily dial in smaller amounts of Toe.

Again, we need to make sure the lasers are pointing true. We can do this by setting plates at the front and rear of the wheels at the same distance from the center of the wheel. These targets should be equal distant from each other at each end when there is zero Toe. Set zero toe with the toe plates, a much more simple task than measuring some amount of Toe and then check the laser to see if the beams are running parallel.

If we are using a four times multiplier, for an 85 inch tire, the diameter of the tire is 27.056". Four times that would be 108.22 inches, or simply 108 and ¼ inch. A Toe of 1/8" would measure a full half inch at the target. It is much easier to see and read, and more accurate than using a tape measure on a toe plate.

**How To Set Toe:**

**Step 1** - Wheel Run-out Check – Check both the front wheels and the rear wheels for Run-out. This means that as the wheel rotates, the outer edge of the tire will wobble slightly. We must compensate for this slight distortion by finding the extreme high spot at a point equal in height to the hub height.

We can simply use a jack stand to hold the tape steady and rotate the tire noting the distance from the stand. Once we locate the high point, we mark it with an arrow and then rotate the tire (be it front or rear) so the arrow is at the top pointing straight up.

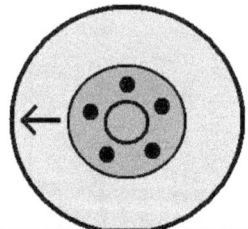 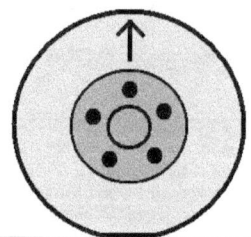

1) Mark the high point at axle height with an arrow.

2) Rotate the tire so that the arrow is pointed straight up.

*Before we set the toe in our race car using toe plates that will rest against the tire sidewalls, we need to find the high point in the sidewall of the tire. There is almost always a high point in a tire and if we place a tape measure or end of a screw driver for instance against the tire sidewall while we rotate the tire, we can visually see the high point. We then mark that point and position it to the top. That way the high point will not point our toe plate at some angle from perpendicular to the axle line and parallel to the wheel.*

**Step 2** - Center the Steering Box – Center the front steering rack. This is done by turning the steering wheel lock to lock and back half the number of turns from full lock in either direction. Once mid-rack (or mid-box with drag link steering) is found, lock the steering shaft against the frame rail with two vise-grip type of pliers.

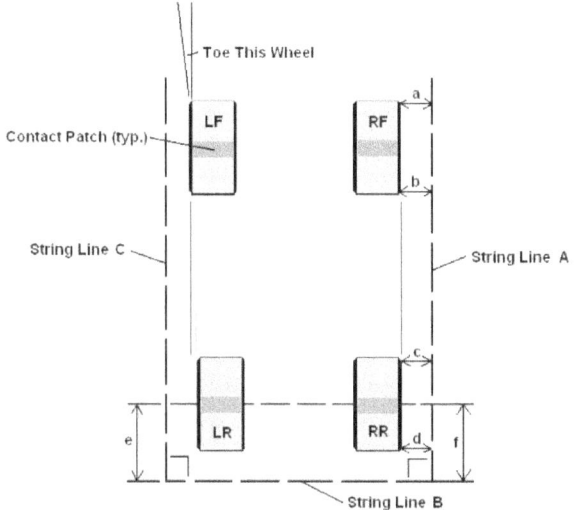

*Another critical point to make sure of when setting toe is to center the steering rack or box, then adjust the tie rods so that both wheels are pointed straight ahead. Then we can add toe-out to the front wheels or toe-in to the rear wheels in a AA-arm suspension system. We can add all of the toe to the left front, right front, or split the toe and add to both the front wheels in a AA-arm front suspension. For the rear AA-arm suspensions, we always split the total amount of toe between the two wheels.*

We want to make sure the steering is centered and the wheels are pointed straight ahead. Once the steering box has been set to center, adjust each tie rod length so that the right and left wheels are pointing straight ahead. With the laser systems, this is done quickly and accurately.

**Step 3** - Setting Front Toe – It is time to set the static toe at the front wheels. We can use toe plates or strings to measure the Toe at the wheel and/or tire sidewall. Remember to be careful, be accurate and do the measurements several times to be sure of the numbers.

If you use lasers, you need to check that the laser is pointed parallel to the wheel and tire. We told you how to do that above. Then just read the measurement from straight ahead and divide by the multiplier of the distance from the center of the wheel.

*In this view, we can see how accurately we can read the toe amounts from straight ahead, being the line the laser is lighting up. When the plates are a distance from the wheel hub by say four times the tire diameter, then an 1/8 inch of toe becomes a distance of ½ inch on the plate.*

**Step 4** – We need to decide how we are going to set the Toe. There are three different ways to do this and it really depends on your preference and the type of race car.

For circle track cars turning only one way, here are some theories. You can either only adjust the left side tie rod to set your toe, leaving the right side alone, or half the toe amount to be split between each side. For 1/8' Toe, the left wheel would be set 1/16" out and the right wheel would be set 1/16" out.

If only setting one wheel for the total Toe you want, we may want to set the left wheel only as opposed to the right wheel because with most circle track cars, the caster split causes the car to pull to the left and driver to steer slightly right when going down the straightaway. Setting the Toe in the left wheel only adds to this pull and the driver won't really notice the effect.

Another way to look at this is to set the Toe with only the right wheel. This way, the steering wants to center left of straight ahead due to the Toe being set this way and that opposed the caster split pull to the right. The driver does not need to steer as much right as before and the car might track much straighter.

After you have set the Toe, roll the car back and forth and recheck the toe setting. Remember that for toe-out, the front measurement on the toe plate will always be more than the rear measurement. If you are using a string or a remote laser to check your toe-out, the opposite is true. The front measurement from the laser/string to the wheel would be less than the rear measurement.

Remember to re-check the Toe after an on-track incident. Small collisions can bend the tie rods and change the Toe. Re-think the Ackermann settings because Ackermann will change the Toe in the car as the wheels are turned.

# Exam - In The Context Of This Lesson:

### Toe Settings In A Race Car Are?

1) Where the wheels are pointed different from in-line with centerline

2) Always with the wheels pointed in

3) Always with the wheels pointed out

4) Not necessary in a straight axle system

### Toe-In In Race Car Design Is?

1) Good for stabilizing front suspensions

2) Helps stabilize rear AA-arm suspensions

3) Where the front of the tires are wider than the rear

4) Important to use in a straight axle system

### Toe-Out In Race Car Design Is?

1) Good for stabilizing front suspensions

2) Helps stabilize rear AA-arm suspensions

3) Where the front of the tires are narrower than the rear

4) Important to use in a straight axle system

### Before We Set Toe, We Always?

1) Center the steering

2) Put the car at ride height

3) Find the run-out in the tires

4) All of the above

### The Use Of Toe Plates Is?

1) As accurate as lasers

2) More accurate than lasers

3) Less accurate than lasers

4) The only way to measure toe

### The Correct Way To Set Toe Is?

1) Right side out, Left side straight ahead

2) Left side out, Right side straight ahead

3) Both sides our equally

4) Any of the above

## Racecar Technology – Level Two
## Lesson Twelve – Driveline Alignment

Drive Line Alignment is one of the basic design features of the car and it can and must be done initially when building, or re-building, a race car in order to save a lot of work later on moving things around. The basic parts of the alignment are the transmission output shaft, the drive shaft itself and the pinion shaft at the rear end on a straight axle rear.

Most of this drive line technology had initially been developed for production cars, but racing applications require a closer look and slightly different approach to this information. The advantages of this knowledge are in the area of reduced power loss and increased component life. To begin with, we need to know what to call the various components related to our driveshaft.

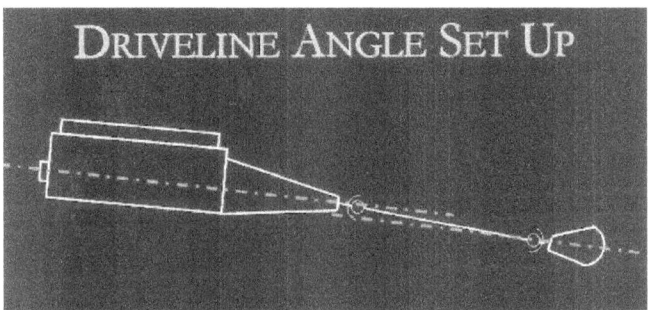

*Proper driveline alignment is when the engine/transmission alignment is parallel to the pinion shaft alignment. If those two are not inline, then there will be an angle created between the transmission and pinion shafts. This angle should be minimal in a racing application.*

Drive Line Terminology – The component names for the driveline parts are as follows:

**Slip Yoke** – this is the yoke at the front of the driveshaft that goes into the transmission and can slip to take up the slack of chassis movement.

**Weld Yoke** – is the yoke at each end of the drive shaft that attaches to the pinion yoke at the rear end and the Slip yoke at the front. These are welded to the shaft tubing.

**Universal Joint Kit or U-Joint (UJ)** – is the actual part that forms the rotational connection between the driveshaft and the transmission and rear end.

**Tubing** – is the tubing between the weld yokes and is usually steel but can be aluminum or carbon fiber.

**Pinion Yoke** – is the yoke that is attached to the pinion shaft at the rear end and accepts the U-Joint.

**Drive Line Angles** – Drive line angle is when the drive shaft tubing is not inline with the transmission output shaft and/or the pinion shaft at the rear end. When the UJ operates with any amount of drive line angles, this creates a problem. The bearings speed up and slow down twice per revolution of the driveshaft. This causes an oscillation in the power train. The more angle that we have, the higher the peaks of oscillation we see and therefore the greater chance of vibration.

Drive line angles are a cause of vibration and power loss. If your race car must have driveline angles from a design standpoint, the angle of the drive shaft to both the transmission output shaft and the pinion shaft should be equal and also opposite. The angle should also be kept to a minimum when at all possible. To get to "equal and opposite", the transmission shaft and the pinion shaft must be parallel.

Drive shaft angles are not only measured from a side view, but also from a top view. Some offset late model circle track and road racing cars can have as much as a 1 ½ inch displacement of the rear of the drive shaft from the front. That equals almost 2 degrees of drive shaft angle at both the transmission and the pinion. So, we can align the drive shaft, from a side view, to zero angle and still have 2 degrees of drive shaft angle present overall.

Strictly for racing applications, near Zero Drive Shaft Angles are encouraged. Based on recent studies and testing, a race car with zero drive shaft angles at the transmission and pinion will not harm the u-joint bearings. The natural vibrations that are produced by the race car will cause the u-joint bearings to rotate enough to stay lubricated and not flat-spot.

In the real world, the drive shaft will never really maintain a zero angle configuration through the diving and rolling of the chassis as we lap the track. Starting at zero means that we will stay very close to zero at all times. If the race car will be at some attitude other than it is at normal ride height, then we must align the driveline with the chassis at the position it will be while we are racing. The teams need to align the drive line for near zero angles with the car placed at the on-track race attitude.

NOTE: For overall consideration in drive line angle design, number one is to keep the driveline angles as low as possible and keep the angles equal and opposite at each end of the drive shaft.

**Controlling Driveline Vibration** – The sources of vibration in our racing driveline, in order of importance – most critical at the top, are:

1) Run-out (similar to wheel runout)

2) Improper Drive line Angles (not equal and opposite)

3) Looseness in the fit of any of the parts (commonly in the U-joints)

4) Unbalanced parts (drive shaft)

5) Component Deflection (shaft wobbles at high speeds)

6) Reaching Critical Speed (a speed too great for the diameter of the tubing)

We will examine some of these causes and see how we can cure many of them just by using better parts that are designed for the racing environment.

If your race car must have drive line angles from a design standpoint, the angle of the drive shaft to both the transmission output shaft and the pinion shaft should be equal and also opposite.

**Yoke Design** – The only attachment design for holding the U-joints that is suitable for all out racing is the strap design. The U-bolt attachment that is commonly used on passenger cars and trucks is no longer considered acceptable for several reasons: 1) it has less strength than the strap design, 2) it may distort the bearing caps if over-torqued, and 3) it grips the cap at three points whereas the strap design grips it in four places which translates to less distortion of the caps.

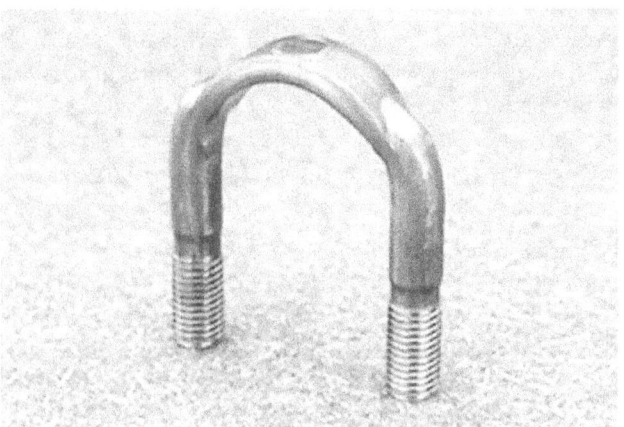

*For racing we should always use parts designed for racing applications. This stock type of U-joint strap is not to be used in racing. There are stronger yokes and straps that will endure the high stress caused by racing.*

**U-Joint Kit Designs** – There are basically two designs of U-joints available. The popular standard zerk designs is used for production vehicles and has, as is obvious, a grease zerk fitting for lubrication. Inherent in this design are hollow shafts that provide the means for the grease to reach the bearings. This hollow design also makes the part weaker than if it were solid.

*In this speedway application, note the heavy yoke and thick strap holding the U-joint in place. And note the large diameter drive shaft tubing. The higher the RPM of the drive shaft, the larger the tubing needs to be in order to prevent vibrations and failure.*

The other design is called the Sealed Design, or Solid U-joint and has no grease fitting and therefore a solid core. This unit has precision seals that keep the lubricant with the bearings while also sealing out dirt. The sealed design is therefore much stronger.

Proper lubrication of the sealed U-joint is simple but can be overdone. We always want to coat the bearings, but not excessively. We also need to fill the trunnion cavity with grease, but not overfill. If too much grease is applied, then when we attach the caps, the pressure from the grease trying to escape will literally blow out the seals and ruin them.

**Adjusting Run-out at the Pinion Yoke** – A major cause of driveline vibrations is when the pinion yoke has a measurable degree of run-out. We can test the run-out after the installation of the U-joint in the pinion yoke by using a dial indicator that is attached to the rear end housing.

*Runout in the yokes can cause vibrations that can damage your driveline system. This tool is used to check for runout in professional applications. You can use a dial indicator mounted to the rear housing and measuring to the end caps. If those are the same distance off of the centerline of the yoke, then the driveshaft is centered on the yoke shaft.*

We measure at both sides of the u-joint and then if the offset is different, we must remove the pinion yoke and rotate it on the pinion shaft to find a position that will index more correctly. This is a very important and necessary step in reducing driveline vibration.

**Failure Mode – Fatigue Crack Propagation** – Drive line components can fail due to a number of reasons. Cracks in the welded yoke or drive shaft will accelerate the failure of the drive line system.

Cracks can occur from the stress of an already weak part or upon construction of the parts due to the heat of welding. It is recommended to weld any drive line part using the TIG welding technique instead of MIG welding.

**Balancing the Drive Assembly** – The only true way to balance a drive shaft is through the use of a Two Plane balancer, or one that simultaneously balances both ends and has the capability to Cross Talk between ends. This allows the equipment to derive a dynamic model of the forces of the imbalance to determine the force vectors involved and formulate a solution that takes into account both ends influence in the overall balance of the shaft.

A simple automotive drive shaft balancer is fine for production automobiles, but for racing, with the very high RPM we experience, we need more precision.

**Snap Ring Failure** – A common failure in our racing driveline is when the retainer snap rings come out of the yoke. There is a simple and easy way to reduce this occurrence by applying a spot of epoxy to the ring. This prevents the ring from collapsing and falling out of the ring groove.

*In this photo, you can see where this team placed a small dab of epoxy in between the ends of the snap ring so they cannot move and allow the ring to come out.*

### How To Align A Drive Line System:

Here is how we can determine what we have, what to do and how to fix it. You will need to put your car up on blocks or boxes so that you can get underneath of it safely. Be sure to block the tires from moving while doing this process. Once the car is secure, we will need to take several measurements.

*You need to align the drive shaft with the car at the attitude it will be while racing. Note the raised left rear corner and compressed right front corner. When this team measured their driveline angles at this attitude, the car needed to be re-aligned. Because the engine was angled down to the rear, the pinion shaft needed to be aligned up to the front to make it parallel to the engine/transmission centerline. When this was tested, the car became vibration free.*

The attitude of the car needs to be the same as it is when running around the race track. So, if you are running on bumps, put the front of the car down to where it would be while on the bumps. For a dirt modified, you could jack up the front on the left side and down on the right just like it would be during the race. The left rear would also be hiked up again like it will race.

Also consider the change in pinion angle if you run a lift arm or pull-bar. Those pieces of equipment will change the pinion angle as they move under acceleration. When power is applied to the drive train, any miss-alignment will cause the most damage. You need to measure your pinion with the rear end at the position it will be under full acceleration.

**How And What To Measure** – First measure your transmission output shaft angle. There is an easy way to do that. Your engine is in line with this part. The valve covers, unless they are of a strange design, will be parallel to the block deck, which is parallel to the driveshaft which is parallel to the transmission shaft. So, just lay an angle finder on the edge of the valve cover, read the angle and write down the number. Also note which way it is inclined, down to the rear or up.

Next crawl under the car and lay your angle finder on top of the driveshaft and measure the driveshaft angle and note the inclination. Then measure the pinion angle and also note the inclination. To get the pinion angle, you might have to use a straight edge placed against the flat part of the flange with the angle finder against that.

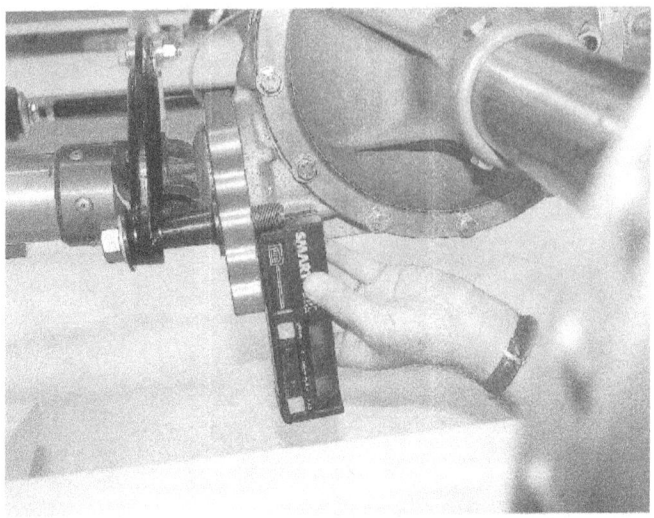

*The first measurements you take are of the pinion shaft, ninety degrees off of this measurement shown, and of the engine/transmission shaft angle. These two should be equal and in the opposite direction (Example: engine up to the rear/pinion down to the front) for proper driveline alignment. We make changes to the engine/transmission and/or pinion angle to make them line up.*

Driveline angles can occur from any view. We have just measured the angles from a side view, but what about the top view? Now we need to look at that scenario.

To measure top view angularity, we need to do some lateral measurements to the center of the tranny shaft and the center of the pinion shaft. If you have a straight rail car and know that the rail is parallel to the centerline of the chassis, this will be easier. If not, you need to setup a string line that is parallel to the centerline.

Measure from the centerline or straight rail to each of the above mentioned shafts using plum bobs or similar devices to make sure the measurements are correct. We need to determine if the two shafts are in line, or if not, how far they are out of line.

For most offset late models, I have found the two to be out of line by 1.0 to 1.5 inches. For a 40 inch long driveshaft that is out of line by 1.5 inches, the driveshaft to pinion and driveshaft to tranny shaft angles would be 2.15 degrees. That is plenty of pinion angle and almost too much by today's standards. But the good thing is that they will be equal and opposite if the rear end is at 90 degrees to the chassis centerline and the engine and transmission is also parallel to the centerline.

*After we have lined up the engine/transmission and pinion shafts, we can then measure the drive shaft to see what the difference in angle is. This is out driveline angle and it should be minimal. If it is over 2.0 degrees, then we need to make changes to the engine angle to reduce the driveline angle. We then re-adjust the pinion angle to match the new engine angle.*

If those conditions are met, then you would not need any side view angles in either the pinion or tranny. So, you could place the tranny shaft, driveshaft and pinion shaft in line with no side view angular difference. That is because the U-joints don't really know which direction they are mis-aligned, only that there are equal and opposite angles in their relationship to the driveshaft.

**How To Adjust The Angles** – Now that we know what the angles are, let's see how we go about making changes. Let's assume that from a top view, the tranny is in line with the pinion. So, we only need to work with the side view angles.

For an example, we have an engine that is downhill to the rear by 6 degrees and the pinion going downhill to the front by 4 degrees. We can reduce the engine angle, but it is impossible to run it uphill to the rear. So we could re-shimmed the motor mounts we get to only 2 degrees downhill to the rear.

Next, it is fairly easy to change the pinion angle because we have a three link rear suspension in our sample car. We then rotate the pinion by extending the length of the third link until it is pointed uphill to the front to an angle that is equal and opposite of the transmission angle, or the same 2 degrees.

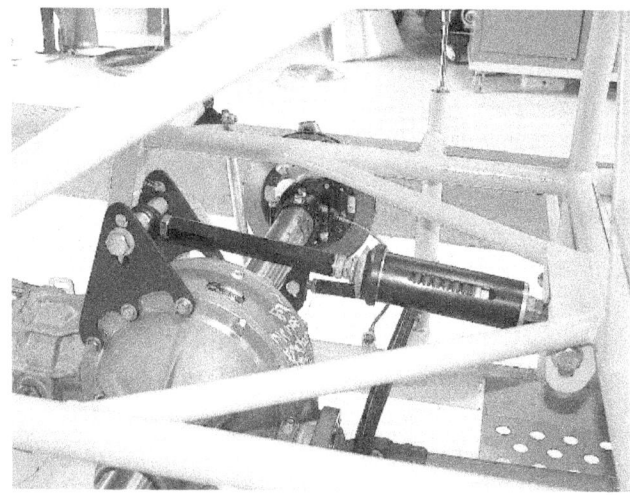

If you have designs different than the solid axle three link suspension, you'll need to get creative when changing the pinion angle. For cars with leaf spring or truck arm rear suspensions, there are wedges made to varying degrees you can use to place between the spring or truck arm and the pad on the axle tube. This will rotate the rear end and pinion to change the angle to the drive shaft.

*Some designs like this lift arm in a dirt late model, promote rotation of the rear end upon acceleration. The forces that drive the car forward also force the arm up at the front to soften the application of power and to redistribute the loads on the rear tires. We need to compensate for this movement and rotation of the rear end when we set our pinion angle.*

For stock four link rear suspensions, you'll need to get even more creative. Since the alignment of the driveline is so important, it is not out of the question to cut and re-weld the suspension links to change the lengths in order to rotate the rear end and change the pinion angle. Just do it in a safe manner. Make good welds or have a professional do it.

**Summary** – The important things to remember are to use drive line parts that are made specifically for racing when possible. Make sure you install the components correctly to reduce drive line vibrations and parts failures.

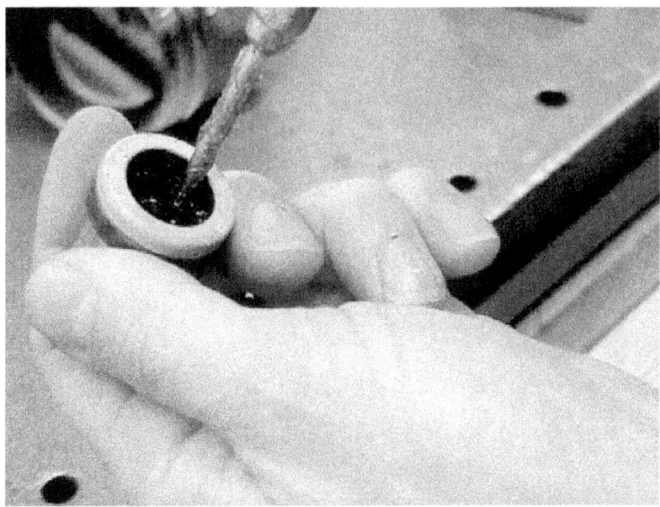

*Lubrication of the U-joint caps must be maintained on a regular basis. Racing U-joints don't have the grease fittings like the OEM U-joints do. Also, be careful to provide enough lubricant but not too much or when the caps are installed, the excess lubricant will blow out the seals.*

*In high speed and high horsepower applications, teams use specially made pinion yokes that are much stronger than stock units. And, they are carefully balanced as indicated by the drill marks on the side that removed weight in order to provide balance.*

Check the alignment of your system and correct any miss-alignment. The drive line must be aligned when the chassis is at the attitude that it will race at, not at normal ride height. Allow for the deflection of pull bars, lift arms, or other rear suspension parts that move under acceleration. Always use a drive shaft that it is up to the task for the intended RPM range you will be running in your type of racing. Periodically check driveline parts for cracks or for signs of a bent or dented shaft.

# Exam - In The Context Of This Lesson:

### Proper Driveline Angles Are When?

1) The pinion is angled 2 degrees down to the front

2) The transmission shaft and pinion shaft are parallel

3) The drive shaft angles at the transmission and pinion are equal and opposite

4) 2 and 3

### Driveline Angles Occur When?

1) From a side view only

2) From a top view only

3) From either a top view or side view

4) When the transmission, driveshaft and pinion line up

### Driveline Vibrations Can Be Caused By?

1) Runout in the pinion yoke

2) A bent driveshaft

3) A miss-aligned driveline

4) All of the above

### Driveline Angles Are Critical When?

1) The car is at ride height

2) The car is under accelerating

3) The car is decelerating

4) Never

### Top View Miss-alignment Is?

1) Never seen in a race car

2) The best way to achieve driveline alignment

3) Most often seen in circle track race cars

4) A way to achieve driveline alignment

5) 3 and 4

### We Can Adjust Driveline Pinion Angles By Using What Method?

1) Adjusting the length of the third link

2) Adjusting the height of the front of a lift arm

3) Using angled wedges for leaf spring or truck arm suspensions

4) All of the above

### The Pinion Angle Must Always Be?

1) Pointed down to the front

2) Pointed up to the front

3) Made level to the ground

4) Any angle that is equal and opposite of the transmission angle

## Lesson Thirteen – Spring Selection for Balance

In RCT Level One, we spoke about what springs are, what they do and a little about how the chassis reacts to spring rates. This Level Two course is all about going to the race car. We are setting up the race car and working with the springs. So, we need to know a little about what our goals are and how springs can help us achieve those goals.

In earlier Lessons we have alluded to the concept and idea of a balanced setup. The term "balance" is used frequently today in broadcast of motorsports events from Nascar racing to Formula One. So, most everyone has accepted the notion that cars that have perfected the dynamic balance part of their engineering are better cars.

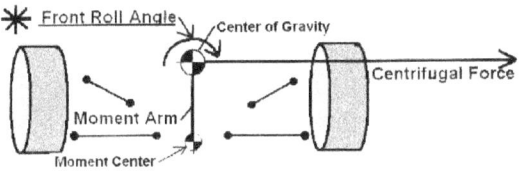

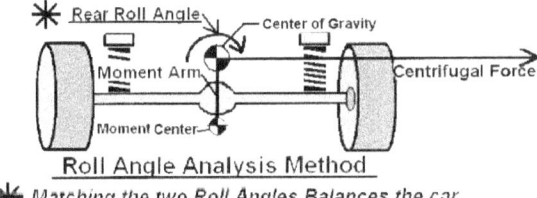
Roll Angle Analysis Method
✱ Matching the two Roll Angles Balances the car

*The concept of a Balanced Setup involves critical elements that we will define and explain in this Lesson. In the most basic terms, it is the front and rear being made to work together to put the maximum and ideal loading on the four tires. This method is a refinement of the older concept of Roll Axis / Roll Stiffness optimization. What RA/RS lacked was the inclusion of influences that affect the "roll stiffness" such as spring split which we will discuss here.*

Now we will present what we hope is a complete explanation as to what constitutes a balanced setup. Just to differentiate, we are not talking about a neutral handling car, we are referring to a dynamic balance where the two suspension systems are in harmony and working together.

The balanced setup has noticeable positive result in race car performance. That is why we are so adamant about getting everyone to fully understand the concept. The following is an excellent definition describing the results of balance and was written by championship racer Larry Bendele. It is a classic and it deserves repeating and so here it is.

"A balanced setup for asphalt has the following characteristics: A perfect handling car allows the driver to have full confidence that he can drive it into the turn as hard as he wants without the slightest worry if it is going to spin out or head for the outside wall. For corner entry, the steering wheel movement is made one time, and doesn't need to be corrected when you approach the corner apex.

"Just prior to the center of the turn, the car rotates slightly (points) without getting loose and begs the driver to give it full throttle. You are able to give it full throttle at the corner apex and keep the hammer down, and the car is willing to run the lower groove on corner exit. However, you can reduce front tire drag (and accelerate more) by purposely steering the car to exit near the outside wall.

"It does the above for 90% of the laps of the feature and you are able to win by several car lengths. Late in the race, you think about taking it easy because of your big lead, however, the car is handling so perfectly and comfortably that it wouldn't be as much fun to drive if you backed it down a notch. So, you smoke the competition."

What Larry describes is what winning race car drivers have sometimes experienced. What we will do in this lesson is teach you how the springs have a major influence in making that happen. In RCT Level Three, we will be setting up real race cars and showing you exactly how springs are used in real situations. For now, we need to come to grips with the process of realizing perfect setup balance.

**New Methodology** - In the mid 1990's a method was developed that represented an advancement related to vehicle dynamic modeling. It involved treating the vehicle as two separate masses, front and rear, each with its own separate suspension system, weights, etc.

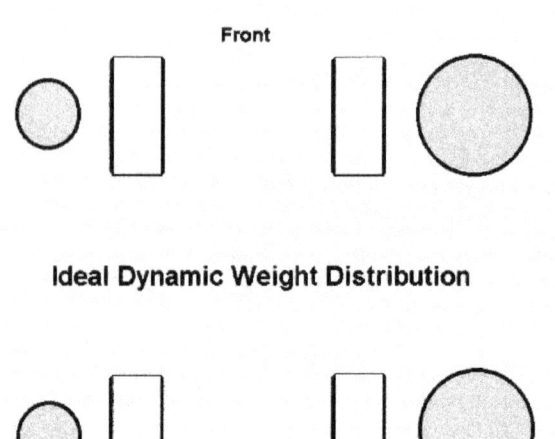

*When we have achieved perfect dynamic balance, our weight distribution after weight transfer has taken place will be as indicated in this sketch. The size of the circles represent loading, the larger the circle the more loading there is on the tire. This is for a car turning left. There will be equal loading on the outside tires, front to rear, and equal loading on the inside tires, front to rear. This provides the same grip level for each axle and also represents a perfectly balanced race car.*

This method made a lot of practical sense because in a stock car we have two "axles", each supporting the weight of each end of the car, with each resisting the lateral forces created by cornering. In physics, we are taught that everything is fluid, or somewhat flexible, and dividing the car into separate halves allows us to analyze each one independently to determine its desire.

There are several critical reasons why a balanced setup is essential to optimum chassis performance. If a cars setup is balanced, then we can accurately predict load transfer. An unbalanced setup redistributes the loads on the four tires in a very unpredictable way. If we cannot match the desires of each end of the car, we then cannot accurately predict the exact amount of load transfer and ultimately how much load ends up on each tire at mid-turn.

We need to have the two sets of tires at each end of the car doing equal amounts of work at mid-turn in order to have a true balance. In most racing series and types of race cars, we will almost never have all four tires doing equal work (having equal loading on each tire) under most current weight rules. If we can setup the car so that after the load transfers in the corners, we have equal working pairs of tires, then we will have a truly balanced setup.

We will also have less (almost non-existent) chassis flex with a balanced setup. Compliance, or flexing of the chassis, cannot occur if we remove the forces that cause this to happen. When the two suspension systems are unbalanced, they are trying to roll to different angles and that causes the chassis to twist.

More importantly, a balanced setup is much more forgiving when the track conditions change or the driver runs different grooves. The speed of the car does not fall of as much, either, as the race goes on and the tires wear. It is the tendency of the balanced setup to maintain a neutral handling characteristic and retained speed after a long run and that helps win races.

**The AA-arm Suspension** - A double A-arm suspension, the type used on the front of most circle track race cars and also on the rear of formula type race cars, has a moment center that represents the bottom of the moment arm. The top of the moment arm is the overall center of gravity of the sprung mass of the car for both the front and rear suspensions and their moment arms.

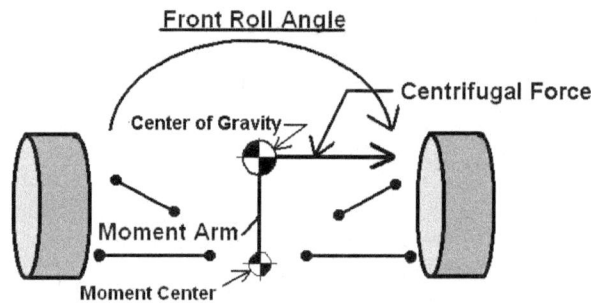

*This sketch represents a AA-arm suspension and how the lateral force acts on the CG. The top of the moment arm is the CG and the bottom is the MC. This force acting through the moment arm tries to roll the chassis. If the front and rear suspensions are trying to roll to the exact same angle, then we have a perfectly balanced setup and we can then calculate the loading on the four tires accurately.*

When the center of gravity/center of mass tries to continue in a straight line as we turn the corner, a lateral force (Centrifugal Force) is exerted on the chassis and that force is resisted by the tires and acts on the center of gravity and resisted by the moment center. If you stick a shovel blade firmly into the ground and then pull on the end of the handle, your arms represent the lateral G-force, the end of the shovel handle is the center of gravity, and the blade at the ground is the moment center.

**The Rear Suspension Dynamics** - The rear suspension is a much different system than the front and we look at it differently. At the front with the double A-arm suspensions, the spring base is felt at the wheels. But a rear solid axle system (and this relates also to a front straight axle car) has a spring base felt on top of the actual springs.

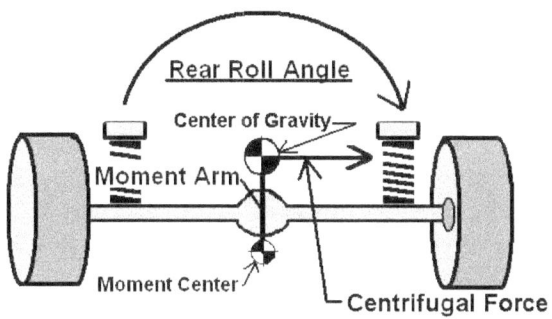

*For a solid axle suspension, the chassis rides on the top of the springs. The spring base is the distance between the top of the two springs for this system, unlike the AA-arm systems where the spring rates are translated out to the tire to create a wheel rate. There is no wheel rate for a solid axle system for the calculation of the roll angle and roll resistance. In this system, the moment arm is like with the AA-arm system, it is the distance between the CG and the MC.*

This dynamic model is not a new concept, but was developed and published some sixty years ago. It is also described and illustrated in the popular book by the late race engineer Mr. Carroll Smith called Tune to Win. So, the width of the top of the springs represents the Spring Base.

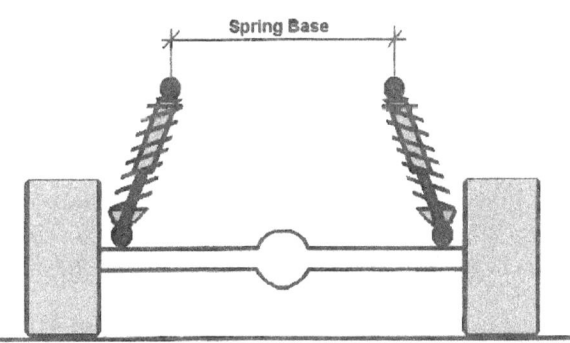

*For a solid axle suspension, as we have discussed, the spring base is the distance between the top of the springs. Even with the springs being angled with the bottom wider than the top, we still consider the spring base at the top where the chassis rests on the springs. The angle of the springs will still influence the rate of the spring "felt" by the chassis, but that ultimate rate will be felt at the top. This is an important concept to understand because with some race cars, it is possible to move the top of the springs in or out from the centerline of the chassis. This can increase or decrease the spring base width and influence the roll angle.*

The rear suspension has a center of gravity of the sprung weight of the car that represents the top of the moment arm just like at the front. The bottom of the moment arm is the moment center created by the lateral locating device known to us by the terms Panhard/J-bar, metric 4-links, leaf springs, or Watts link.

These four devices comprise the majority of lateral restraint systems used for straight axle suspensions in race cars. Each restraint system has its own moment center that is the bottom of the rear moment arm.

**Two Forces at Work** – The lateral force is not the only force acting on a race car here on Earth. Again, the moment arm is much like a pry-bar or shovel handle. The CG is equivalent to the end of the bar we hold on to, and the roll centers are the opposite end, which is the object we are trying to move. The longer the bar (moment arm) the more leverage we have and the easier we can move the object, or in this case, roll the car. Now here is what else is happening.

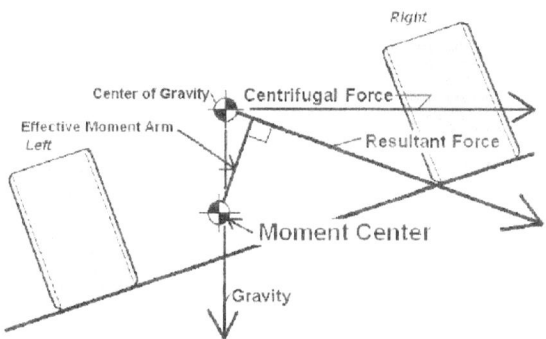

*In every race car, there are two forces acting on the chassis. There is the ever present Gravity and then there is the lateral force we call Centrifugal force which is acting parallel to the path of the car going through the turns. In almost every case, this path is parallel to the earths surface, or to be more specific, perpendicular to a line running from the CG to the center of the earth. In physics, when two forces act on a body, they combine into a Resultant Force which has a magnitude and direction.*

In a race car we really have two forces at work being applied to the center of gravity. One is the lateral force of cornering we described as the centrifugal force, and the other is Gravity. We cannot forget about this basic force that is acting on the chassis at all times, even when at rest. The two forces combine into one resultant force for which we can calculate a magnitude and direction.

By looking at the direction of the resultant force in the sketch, we can see the true picture of how a combination of the two forces will react in the front suspension through the Center of Gravity and be resisted by the moment center. The Effective Moment Arm length or Combined Force Vector is the result of the direction of the combined force and the location of the moment center.

**Effective Moment Arm** - The Moment Arm length is the vertical distance between the CG height and the MC height. The amount of lateral force and the length of the effective moment arm are contributing factors that help determine exactly what angle the front of our car desires roll to. If the front end of our car was not rigidly connected to the rear it would roll to a certain angle and a predictable amount of load would transfer from the left front tire onto the right front tire. This is the Front Roll Angle.

But we know that the two ends of the race car are connected with a rigid chassis that resists twisting. The two ends of the car are connected and unless we know for sure that their desires are matched, then we cannot accurately predict the load transfer at each end of the car.

**Primary Influences On Roll Angles** – The following are the primary influences that affect the magnitude of the front and rear roll angles. They are:

1) The weight of the sprung mass of the portion of the car supported by the suspension (scale weights at each end minus the weight of the un-sprung components),

2) The height of the sprung mass Center of Gravity,

3) The height of the Moment Centers front and rear,

4) The magnitude of the lateral force (measured as G-force),

5) The overall spring stiffness as well as side to side spring split (i.e., right rear stiffer than the left rear spring, etc.) at each end,

6) The width of the spring base, which is the distance between the centers of the top of the springs for a solid axle suspension and the track width of the tires for a AA-arm suspension,

7) The stiffness of the sway bar (if used) which has an effect of anti-roll and must be taken into account. The larger the diameter of the bar in this system, the more resistance there is to roll,

8) And the Track Banking angle.

Now that we know what influences the roll desires of each suspension system, the key to creating a balanced setup is to change spring rates, sway bar, etc. so that we can find the combination that will cause each end of the car to WANT to roll to the same angle. We can detect when our car is balanced by observing tire temperatures as well as tire wear.

In order to set up our cars, we would need to be able to change components, if necessary, to balance the desired roll angles of the front and rear of the car. These changeable components include the spring rates, spring split, the moment center heights front and rear, and if used, the sway bar sizes and arm lengths.

**A Balanced Setup** – A dynamically balanced setup is where both ends are working together and WANTING to roll to the exact same angle. Some teams will experiment and sometimes find a really good setup and they might not know how they got there or why it worked so well. Often the performance could not be duplicated when they built, bought or otherwise acquired a new car because of differences in moment arm locations or other differences that affect the roll angles.

The Balanced Setup concept is one that means we should work to match the front to rear desires until our setup is balanced. The "desire' can be thought of as the roll angles that each end of the car would want to achieve if that end were independent and not attached to the other end.

For a front AA-arm / rear solid axle car, the front roll angle is not easily adjustable within the range of spring stiffness we usually want to work within. So, to balance the setup in these cars, we need to work with the rear of the car and make changes there to match what the front wants to do.

Our choices for adjustments include making spring changes for both stiffness and spring split for circle track cars, raising or lowering the rear moment center, and that is it. So, spring split is the primary way to get close to the balance we need and the panhard bar is the fine tuning device to use once we get close.

For road racing cars with AA-arm front and rear suspensions, we work with overall spring stiffness combined with the anti-roll bar stiffness. We can also tune the MC height for each system and we will get into more on that in RCT Level Three.

**How to Select the Right Springs** - The desires of both ends of the car can be matched through the use of many combinations of spring rates, spring split amounts, and MC heights. So, we need to make a decision as to how soft or stiff our spring rates need to be for the track we will be running on. The general rule is to stiffen the spring rates for high banked and faster tracks and soften them for lower banked tracks with less speed and traction.

For road racing cars, we run equal spring rates on each side of the car at each end. What we call spring split, or running different rate springs on each side does not apply to road racing cars for the most part. So, for those cars, spring stiffness, MC height and sway bar rates front and rear, are the deciding factors in determining the roll stiffness and ultimately the roll angles for the front and rear.

**Circle Track Race Cars** - Spring split at the front can enhance the transitional parts of the track. On flatter tracks, a softer RF spring than the LF spring, can improve corner entry. This is because as we brake, a softer RF spring can start the chassis rolling in the same direction that it will roll to in the middle of the turns. On medium and higher banked tracks, there is no need to run a softer RF spring and we would either even them up or run a stiffer RF spring on tracks banked 14 degrees or higher.

The same is basically true for the rear of the car. A softer right rear spring for lower banked tracks for cars using the Metric four link works where improved bite off the corner is usually needed. In addition, those cars inherently have a very high rear Moment Center and this form of spring split will help create a rear roll angle to match the front.

In modern day circle track racing, it is common to run a stiffer RR spring rate to try and match the low front roll angle created by using stiff bump stops or bump springs. The force needed to do that in the RR can be pre-loaded into a softer rate of spring so that the RR corner will compress more on exit. We'll describe in detail how that happens and how to do it in RCT Level Three.

On the higher banked tracks, we can run the same rate for the rear spring, or use a higher RR spring rate than the LR spring rate to create an anti-roll effect. To help reduce the roll tendencies for the Bump setups where the front has a low desire to roll when riding on the bumps, a much stiffer RR spring rate will help reduce most of the roll tendencies in the rear. Or as we have stated, a softer spring rate pre-loaded to a sufficient force to resist roll.

Whichever type of race track you run and whether you run dirt or asphalt, attention must be paid to the relationship between the front suspensions desires and the rear suspension desires. When both ends are close to wanting to roll to the same angle in the turns (more easily referred to as desired roll angle), then the more work all four tires will be doing and the faster and more consistent your car will be.

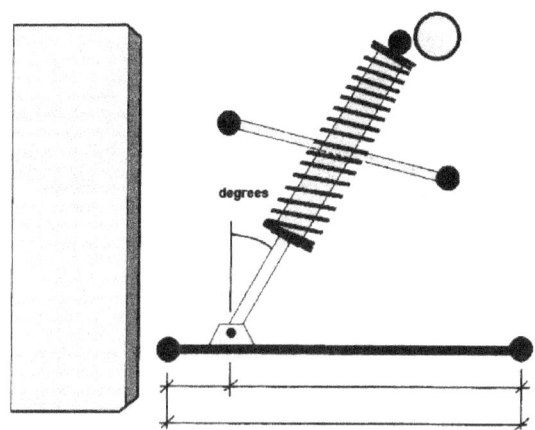

*The front spring rate is translated out to the wheel for a Wheel Rate in this coil-over spring system. We need to find the motion ratio and spring angle to find the Wheel Rate for this application in order to find the roll angle.*

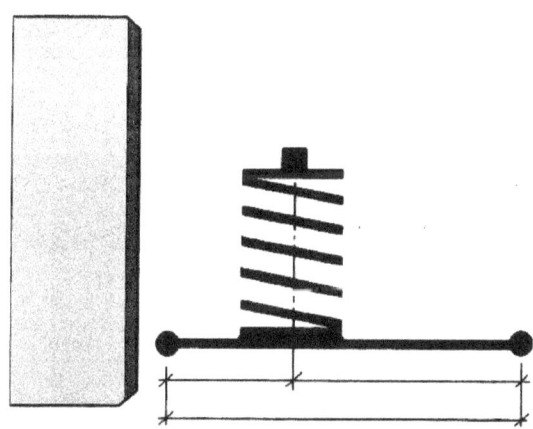

*In a way similar to the coil-over system, the stock type of big spring suspension has a motion ratio that must be used to find the wheel rate. The wheel rates and the track width (spring base) are determining factors in finding the roll angle.*

**What Affects Roll Angles** – The following are general rules concerning roll angles and can be considered when wanting to know how to create larger or smaller roll angles.

*Spring Rates* – A higher overall average spring rate at the front or rear results in a lower roll angle. A lower overall average spring rate at the front or rear results in a higher roll angle.

*Spring Split* – A spring split with the inside spring softer than the outside spring results in a reduced roll angle. A spring split with the outside spring softer than the inside spring results in a greater roll angle. And, increasing the inside spring rate results in a greater roll angle. Decreasing the inside spring rate results in a reduced roll angle. Conversely, decreasing the outside spring rate results in a greater roll angle and increasing the outside spring rate (or spring force) results in a reduced roll angle.

*Spring Base* – In a solid axle suspension, a wider spring base results in a reduced roll angle. A narrower spring base creates a greater roll angle. Remember that the chassis feels the spring base at the top of the springs, not at the bottom. It basically rides on the tops of the springs.

*Moment Center* – A higher moment center at the front or rear will result in a reduced roll angle. The moment arm becomes shorter and has less overturning moment as a result. Conversely, a lower roll center at the front or rear results in a greater roll angle.

*Sway Bars* – Changing to a larger diameter sway bar, thicker wall, shorter length sway bar, or shorter sway bar arms, results in a reduced roll angle. Conversely, changing to a smaller sway bar, thinner wall, longer length sway bar, or longer sway bar arms, creates a greater roll angle.

*Lateral G-force* – A higher lateral G-force results in a greater roll angle, and a lower lateral G-force results in a reduced roll angle.

*Center of Gravity Height* – A higher center of gravity results in a greater roll angle. A lower center of gravity creates a lower roll angle.

*Track Banking Angle* – If the G-forces remain the same, the following apply. A higher banking angle results in a reduced roll angle. A lower banking angle results in a greater roll angle. However, a higher banking angle usually results in the race car being able to achieve a higher speed and higher Lateral G-force and therefore must be considered in the influences on roll angle.

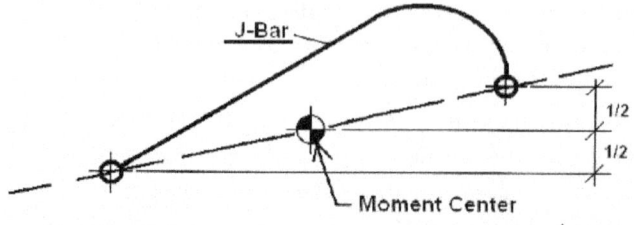

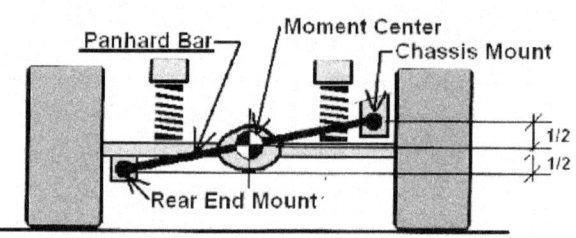

*The rear moment center height is the average height of the center of the bolts connecting the ends of the panhard, or J-bar. When we raise or lower the ends of the bar, we raise and lower the rear moment center affecting the roll angle of that end of the car. As the car rolls through the turns, if the outside end of the panhard bar is mounted to the chassis, it will move down causing the rear MC to also move down. If the car squats on exit, then the chassis end of the panhard bar will move down and with it the rear MC. These are all things to take into account when working out our setups.*

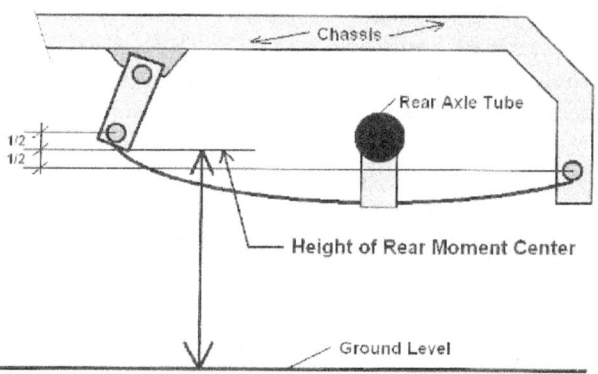

*The height of the rear MC for a leaf spring suspension is the average height of the two mounting bolts on the leaf spring. As the chassis rolls through the turns, these points move down on the outside leaf just like the chassis mount of a panhard bar does lowering the rear MC.*

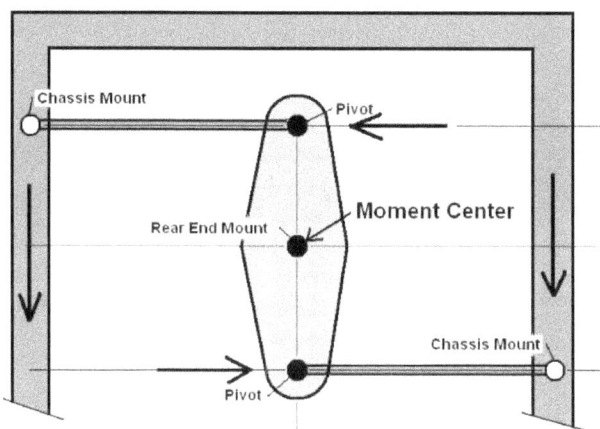

*The Watts link system, used mostly with road racing cars using solid axle rear suspensions, is a system that maintains the moment center height through chassis roll. The point of resistance to lateral forces is at the vertical center of the rear end and mounted to the rear end. The other notable feature of this link is that it does not influence the chassis to move side to side as the car dives and rolls. There is no rear steer due to lateral movement of the chassis with this system.*

**How To Tell** - Tire temperatures and tire wear tell us a lot about how our car is working. Equal front to rear tire temperature averages indicate a neutral car, though that car might not be well balanced. Equal temperatures and/or tire wear on each side of the car is an indicator of a balanced setup. When the LF and LR tires are doing equal work, and the right side tires are doing equal work, you will have found that balanced setup.

Just try to remember which components affect the roll tendencies of each end, and to what extent, and making those adjustments will be quicker and more accurate in your search for that winning setup.

**Aero Balance Influence** – Although this discussion is all about mechanical balance and how we go about achieving that, I need to put in a short mention of aero balance. If we truly achieve mechanical balance and our wheel loads are ideal, then where does aero downforce fit in?

If we end up with a lot of front aero downforce and very little rear downforce, then obviously the car will have more grip on the front tires due to having more loading. We need to be careful when we analyze our aero package that we don't overwhelm one end by creating too much aero downforce on the other end. We'll talk much more about that in RCT Level Three.

**Summary** - All race teams need to know certain basic information about their cars so that the foundation of a balanced setup can be laid. Learn to recognize the tendencies that make a car's suspension want to roll more or less and try to match the two ends of the car for a more balanced setup.

Know your front geometry settings, especially the Moment Center height and camber change characteristics. Be prepared to adjust the weight distribution (cross weight distribution) as you make changes in the direction of a balanced setup.

In RCT Level Three, we will be using this information to setup an actual race car. We will demonstrate how we can perfect corner entry and exit performance without changing the roll angles in the suspensions. What you have read and learned in this Lesson is among the most important pieces of race car technology information you will ever learn.

# Exam - In The Context Of This Lesson:

**Dynamic Balance Is When?**

1) The car is neutral in handling

2) The front and rear weights are the same

3) The two suspensions are working together

4) The inside and outside tire loadings are equal

**Ideal Roll Angle Is When?**

1) The chassis does not roll

2) The front rolls more than the rear

3) The front rolls less than the rear

4) Any roll angle where the front and rear are equal

**Roll Axis/Roll Stiffness and Roll Angle Methods Are The Same?**

1) True

2) False

**Matched Roll Angles Create What?**

1) Ideal weight distribution

2) A faster turn speed

3) More overall grip

4) Better tire wear

5) All of the above

**The Bottom Of The Moment Arm Is?**

1) The ground

2) The bottom of the springs

3) The moment center height

4) The tire contact patches

**The Top Of The Moment Arm Is?**

1) The top of the springs

2) The upper ball joints

3) The upper chassis mounts

4) The center of gravity of the sprung mass

**Name The Two Forces Acting On The Race Car?**

1) The jacking force

2) The overturning moment

3) The force of gravity

4) Centrifugal lateral force

5) 3 and 4

**Spring Base Is Measured At?**

1) The bottom of the springs

2) The height of the springs

3) The top of the springs for a solid axle

4) The tire contact patches for a AA-arm suspension

5) 3 and 4

**Which One Is A Primary Influence On Roll Angles?**

1) The Center of Gravity height

2) The spring stiffness

3) The spring base width

4) The lateral G-force magnitude

5) All of the above

**Which Will Increase The Roll Angle?**

1) Raising the moment center

2) Lowering the CG

3) Softening the RR spring rate

4) Softening the LR spring rate

## Racecar Technology – Level Two
## Lesson Fourteen – Sway / Anti Roll Bars

A sway bar, or what is sometimes called an anti-roll bar is a device that resists chassis roll. The device can be constructed in many different ways and some of the components may not be "bars" at all. They just have to produce resistance to the rolling action of the chassis.

In this Lesson, we will tell you how to rate your sway bar the accurate way. We won't rely on published data or simple calculators and we will tell you why you shouldn't do that. Once we know our installed sway bar rate, we can then use that information to setup the car. We will be doing that in the RCT Level Three course.

*As we prepare our test car for finding the true rate of the sway bars at both the wheel and at the ride spring placement, we install solid links in place of the spring/shock combo. This Method Number One is both easy to do and very effective and accurate. We also can measure the actual force the sway bar exerts on the suspension at mid-turn using the Gale Force Sway Bar force measuring tool.*

As a chassis rolls, the inside (of the turns) springs decompress and the outside springs compress as the car rolls towards the outside of the turn. This is an over-simplification of what occurs, but gets you thinking in the right direction. In reality, the outside springs compress more than the inside springs in most cases and that results in chassis roll.

Chassis roll in and of itself is not a bad thing, but the trend in racing is to limit the amount of roll in a race vehicle. Of course, some forms of racing need the cars to roll in order to facilitate weight transfer and the chassis compliance that is needed. The racing I refer to is dirt racing.

Dirt cars just roll more than other forms of racing and most dirt cars do not even have a sway bar installed. So, we can assume from this point on that we are talking about asphalt race cars including circle track and road racing sedan and formula types of cars. First let me give you a little information about sway bars.

*On the formula type of race car, the systems are necessarily made very compact and the sway bars designs can be varied and somewhat odd. With the links attached to rockers that join the sway bar and the spring running at odd angles from a top view and attached to bendable blade arms at the ends of the bar, we can count many variables and non-linear functions as the car rolls. Calculating the influence of the sway bar becomes almost impossible without the wheel rate measurement method we describe here. The two gray bars are the links from the sway bar arms to the pushrod rockers attached to the springs. The sway bar is to the right and the springs are to the left.*

*The blades on this prototype cars sway bar can be turned to provide varying amounts of resistance to bending. This combines with the twisting of the tubing to produce many combinations of sway bar rate. We would need to test each setting of the blades, different blade thicknesses and different sized sway bars and note those rates for future use.*

**Items That Affect The Rate of The Sway Bar –** There are several things that affect the rate of a sway bar. The bar itself may have a stiffness, but it does not have a rate. When we attach an arm to it, we can then install it and when the car rolls, it twists in resistance to the motion of roll and then has a rate. We need to know the installed rate of the sway bar related to wheel rate and/or ride spring rate. More on that later.

So, the sway bar resists chassis roll and with some of the more popular setups, minimizing and/or eliminating chassis roll is a desired effect helped along by using sway bars. The following are the seven most common variables that affect the amount of work the bar does and there could be more in some applications.

- The length of the portion of the bar that will twist affects its rate. The longer this portion is, the softer the rate.

- The outer diameter of the bar affects its rate. The larger the diameter, the stiffer it will be.

- Hollow bars are less stiff than solid bars, so the wall thickness (or hole size) affects the rate. Thinner walls are softer rates.

- The hardness of the steel used for the bar affects the rate. There is a modulus of elasticity for each type of steel that relates to the mixture of metals and the hardening process used in the manufacturing of the steel. This is the part we cannot absolutely know because it is almost never published.

- The sway bar arm length affects the rate. The longer the arm, the softer the bar rate.

- The arm material can affect the rate. If the arm bends under load, the rate will be softer.

- The installation ratio affects how the bar rate relates to the wheel rate and/or ride spring rate.

It is easy to see that calculating a sway bar rate and knowing how much work it does is very difficult at best. That is why we should measure the rate and force as it is when the bar is installed in the car with all of the motion ratios, arm lengths, and other variable taken into consideration. There just is no other way to accurately do it and be able to use the information for the actual setup.

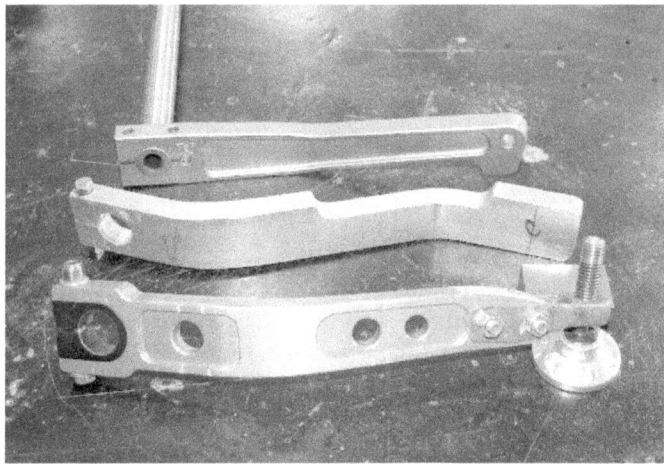

*In the stock car world, sway bars can sometimes be larger for some uses. If the type of racing requires the use of stiffer bar rates, then stiffer arms are needed to connect the bar to the lower control arms. The old-style aluminum arms are just not strong enough and will bend, greatly influencing the overall rate and influence of the bar. And, the rate will not be linear. We proved that in tests done almost twenty years ago. The steel arm in the rear of the photo is best because it flexes very little.*

It is important to know your sway bar rates in order to translate that information to the setup you desire. There are problems associated with looking up the rate in a table or other publication or doing a calculation because of several factors that affect the working rate of the bar. All of the variables are taken into consideration when we rate the bar after it is installed in the race car.

There are several methods that take into account all of the variables to provide the true working rate of any sway bar. The first one has been used successfully in both circle track cars as well as formula prototype race cars where the sway bar system is very complicated.

**Method Number One** - Giving credit where credit is due, I came across this method when it was presented by James Hakewill on his website. With this method, your team can go through the process of measuring the rates of all of the combinations of sway bar sizes and settings for the more complicated sway bars.

The method works for any type and installation of sway bar from stock car types to formula types including Formula One and Indy cars. The method rates the sway bar system and provides a wheel rate of the bar. The wheel rate can be used for many different setup functions. The following is a detailed account of how you can rate your sway bar. This simple yet effective method takes into account all of the variables and any we might not have mentioned. Here is how it works.

**Overall Principle of the Method** — What we are going to do is cause the sway bar assembly to support the weight of one corner of the car. Once we do that, we can accurately calculate the sway bar rate as a wheel rate and then through the motion ratio, as if it were a ride spring so we can know how our sway bar affects our car.

We will use the Right Front corner to rate the bar in this super late model circle track car. The process is the same for other types of race car. Sitting at ride height, the RF tire supports a portion of the total weight of the car. If we can cause the sway bar to act as a spring holding up that corner, then there will be a deflection of the wheel that we can measure and convert to a spring rate at the wheel in pounds per inch.

Once we find the wheel rate of the bar, we will convert that rate back to the ride spring mount to get the rate as a ride spring. From there we can use that information to help setup the car. I think that now you can see where the sway bar rated at the end of the arm is of little use in the real world.

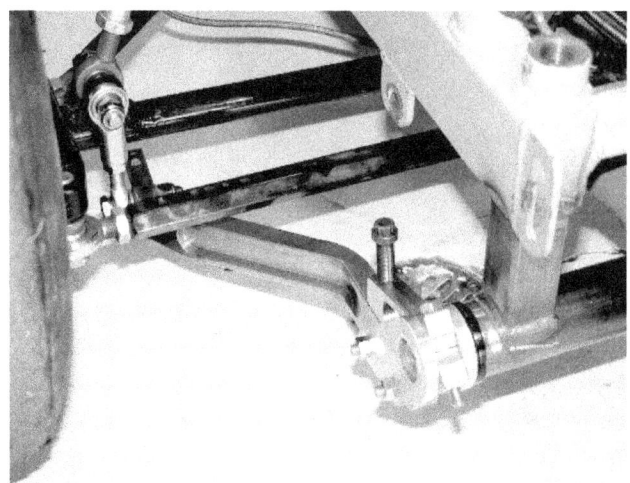

*This sway bar arm attachment to the sway bar itself is easily adjustable. For our testing, we set the bar neutral to find just the bar rate. Then we apply the number of turns of pre-load we would use on the race track to find the sway bar force that combines with the spring and bump force at mid-turn. The sway bar force is added to the spring and bump forces to match the force needed to support the right front suspension at mid-turn.*

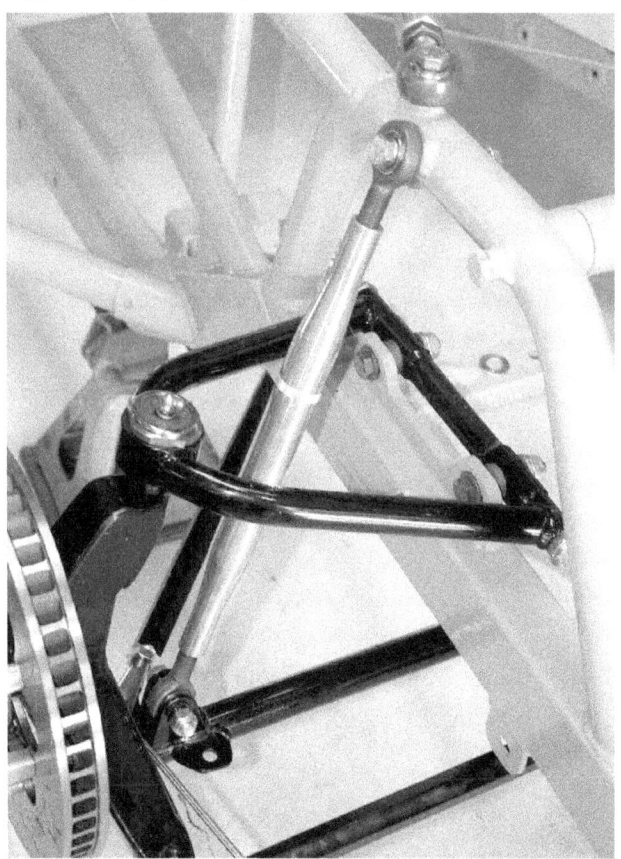

*You will need to install solid links at all four corners of the car in place of the shocks. Make sure your link is strong enough to support the weight of that corner. Using opposite threaded heims helps us to adjust the lengths easily. We will read the loading on the tire over the scales at normal ride height and also the unsprung weight of the wheel, control arms, etc.*

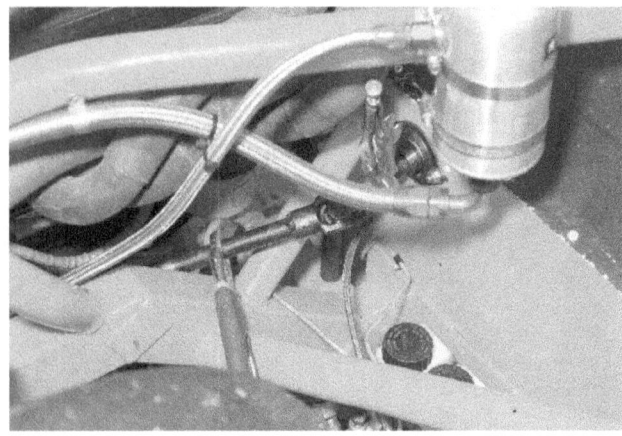

Before taking any measurements or weighing the car, lock the steering shaft with the wheels pointed straight ahead. Movement of the steering and wheels will negatively affect the results. We used opposing locking pliers to hold the steering shaft.

You will need to support the rear of the car at one point in the middle between the two rear wheels. We used a section of large angle iron placed on the cross bar behind the rear end. Remove the rear wheels so that there is room for movement. Then lower the rear to the normal ride height plus any offset such at the height of the scales in this case.

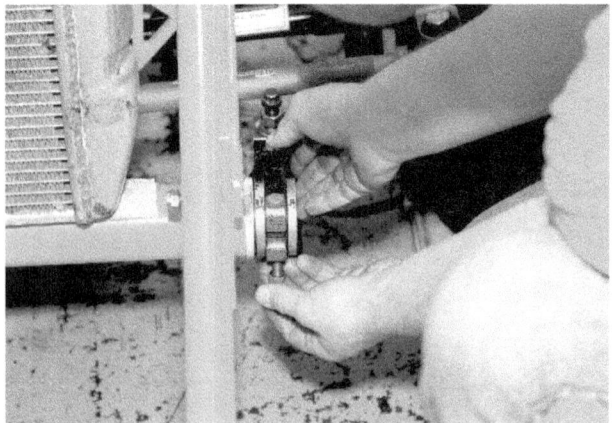

When measuring the static loading on the RF scale with the solid link installed, make sure the bar arm is loose. Then, before getting ready to support the car with the sway bar, snug the left arm up against the lower control arm, but do not pre-load the bar. Then you can find the sway bar wheel rate. If you are measuring mid-turn force, go ahead and pre-load the sway bar just as you would for racing because the pre-load will add force.

Once we have caused the sway bar to support the weight of the RF corner, we need to place thicknesses of wood under the tire in order to shim the tire up so that the chassis ride height remains at the previous static ride height. Remember that when we weigh the un-sprung components, we need to add the weight of those wood shims to the un-sprung weight since they will be a part of the wheel assembly when we supported the car by the sway bar.

*When taking your measurements, take your time. Get down at right angles to the tape and look at it several times to be sure you get the correct reading. Read the tape to 1/32nd of an inch for best results. A small error can add up to a lot of sway bar rate difference.*

*When rating the sway bar, it is a good time to check the connection between the end of the sway bar arm and the control arm. This link needs to be perpendicular to the sway bar arm and the control arm.*

**Getting Started** – We start off by taking the following steps in preparation to take the measurements we will need to determine the sway bar wheel rate. This routine is fairly easy and fun to do. It may take some time, but the results will last forever.

• Install the desired bar and arms and leave the connections loose for now. For stock car installations, make sure that the link attaching the bar to the right control arm is perpendicular to both the sway bar arm and the lower control arm. This is very important and can affect the motion of the bar causing it to bind and be artificially stiff.

• Remove the springs and support both of the front corners of the car with solid links in place of the shocks. Adjust the length of the links to be the exact same length as the shock/springs were at normal ride height. Also make sure that the material used for the links is strong enough to support the weight of the car. We usually use steel tubing with nuts welded onto the ends and heims attached with opposite threading so we can easily adjust the length.

• Support the rear of the car at one point at the center of the frame on a jack stand, or similar, and use a piece of angle iron placed on top of the stand. Set the height to be the same height for the chassis as it is at normal ride height. You will need to remove the rear wheels to provide room for chassis movement.

• Place the front wheels onto the scales and record the RF weight reading.

• Raise the frame at the RF with a jack and remove the link. With the chassis still supported, lower the RF wheel/tire onto the scale and record the weight of the wheel assembly. This is the RF un-sprung weight. Move the assembly up and back onto the scales several times to take the friction out of the pivots. Also weigh each individual shim you will use later on to space under the tire. When we know which shims we need to use, we can add the weight of those shims to the un-sprung weight.

• While the link is out and the load is off the RF wheel, we can measure the shock motion ratio. Measure the shock length (the distance from center of bolt to center of bolt of the mounting bolts for the shock with the tire resting on the scale pad with no load on it.

• Place a thickness of wood or other material under the RF tire. Re-measure the shock length. Divide the amount the "shock" moved by the thickness of the material placed under the tire. Example: if the material was 2.0 inches thick and the shock moved 1.5 inches, then our motion ratio would be 0.750 (1.5/2.0 = 0.750). We will use this number later on.

• Remove the block, put the link back in place, remove the jack and let the front tires again rest on the scales. Snug up the sway bar arm being careful not to pre-load the bar. Record the RF ride height from the frame to the floor. We will need to keep this ride height the same throughout the process.

• Record the weight on the RF scale pad to make sure it is the same as before. Now jack up the RF corner carefully and remove the link. Let the RF corner back down onto the scale so that the sway bar is supporting that corners weight. The frame will be lower and we need to return it to the previous ride height.

• Place the shims under the RF tire so that the chassis will be back to the same original ride height. This is so that no load transfer takes place from chassis roll. The RF scale weight will now include the weight of the shims.

• Now that we know the weight of the shims, subtract the combined RF un-sprung weight, plus the weight of the shim, from the total RF weight to find the RF Sprung Weight that the bar is now supporting. Record that number.

• Measure the new, compressed "shock length". Find the difference between the new shock length and the original static shock length and divide that number by the RF shock motion ratio (0.75) that we found before. This tells us how far the wheel moved vertically. Record that number.

• Divide the sprung weight of the RF corner by the amount the wheel moved to find the Sway Bar Wheel Rate. Example: if our RF sprung weight were 600 pounds and our wheel moved 2.0 inch, then the sway bar wheel rate would be 300 lb./in (600 / 2.0 = 300).

**Finding The Bar Rate** - We now have the wheel rate of the installed sway bar assembly. This is not the last step in the process. We need to know the bar rate as if it were a ride spring. Then we can add that rate to the ride spring rate to determine total ride spring rate. Here is how we do that.

• We already know the motion ratio of the wheel to the ride spring. In this example it is 0.750. The spring moves 0.750 the amount the wheel moves. So, we need to square that number. 0.750 times 0.750 = 0.5625. Our dynamic motion ratio is 0.5625.

*Instructors Note:* To find spring rate from the wheel rate, or the wheel rate from the ride spring rate, we use the motion ratio squared. This is the law of motion ratios as a part of the study of dynamics and forces.

• To find the sway bar rate as a ride spring rate, we divide the Sway Bar Wheel Rate by the sway bar motion ratio squared (0.5625). Example: if we have a sway bar wheel rate of 300 lb./in. and a sway bar dynamic motion ratio of 0.5625, we calculate the sway bar rate as a ride spring rate to be, 300 / 0.5625 = 533 lb./in. The work the sway bar does is like installing a 533 pounds per inch stiffer ride spring.

**What This Tells Us** – Once we find the bar rate, we can relate that to our setup. You might be using a bar that is much stiffer, or softer, than needed for your intended use. In any routine, we can only expect the results to be as accurate as the effort we put into it. We need to make sure all of the steps are done correctly and that no other influences affect the results.

Make sure you zero your scales before taking any readings and lock the steering shaft so the wheels don't turn. This could affect the results. Also take your measurements a couple of times to make sure you read the tape correctly. Take your time in all phases of the process and keep good notes.

**Method Number Two** – This next method will let you rate the sway bar as a force in the same way we rate a spring as to force. Once we know the travel of the wheel and spring in our system, on our race track at mid-turn, we can measure the force the sway bar is adding to the spring/bump combination. Here are the steps.

*This shows the Gale Force sway bar force measuring tool installed on this modified at the right front. With this tool, we can move the suspension up to the same travel as it shows at mid-turn and then measure the force of the sway bar only. We have removed the ride spring and placed the tool in place of the ride spring. The force in pounds is the exact work the sway bar is doing. If we know the ride spring force we will need for this car, we would set the ride spring to the total needed force minus the sway bar force.*

• We first use a Gale Force Sway Bar force measuring tool. Leave the sway bar arm loose for now.

• Mount the GF rig into where the ride spring/shock would mount, adjust it to the correct length equal to normal ride height, and then zero the reading.

• Now set the sway bar links to where they are for on-track racing. If the sway bar is set to neutral, or preloaded some amount of turns, we just duplicate where it is for racing.

- Now jack up the wheel using the GF rig and the included hydraulic pump until we reach a shock length equal to what we read for on-track movement at mid-turn. We can then read the force on the digital readout to know what force the sway bar is adding to the ride spring and bump forces to support that corner of the car at mid-turn.

**What This Is Used For** – If we know that the force needed at the ride spring, to hold up the RF corner of the car at midturn is 2400 pounds, then if we know the sway bar force, we can set our ride spring, plus any bump device, to a force that is 2400 pounds minus the sway bar force. This number will be the net force the ride spring plus bumps will need to produce to support this corner of the car at mid-turn.

If the sway bar force were 533 pounds like the first example, then our spring/bump combo force should be 2400 minus 533 = 1,867 pounds. Now we can accurately set our ride spring height and bump spacing to generate 1,867 pounds of force at the mid-turn travel amount.

The overall use for these forces and how we calculate the required force will all be covered in RCT Level Three in much more detail. This is a very important and necessary part of race car setup and applies to all race cars.

**Old School Method** – The old school methods we have used in the past to rate a sway bar involved using either a scale to read the rate at the end of the sway bar arm, or a formula to calculate the rate at the end of the arm. Both of these methods provide a number that is hard to relate to our setup, but could be used to sort out different sway bars and arm combinations.

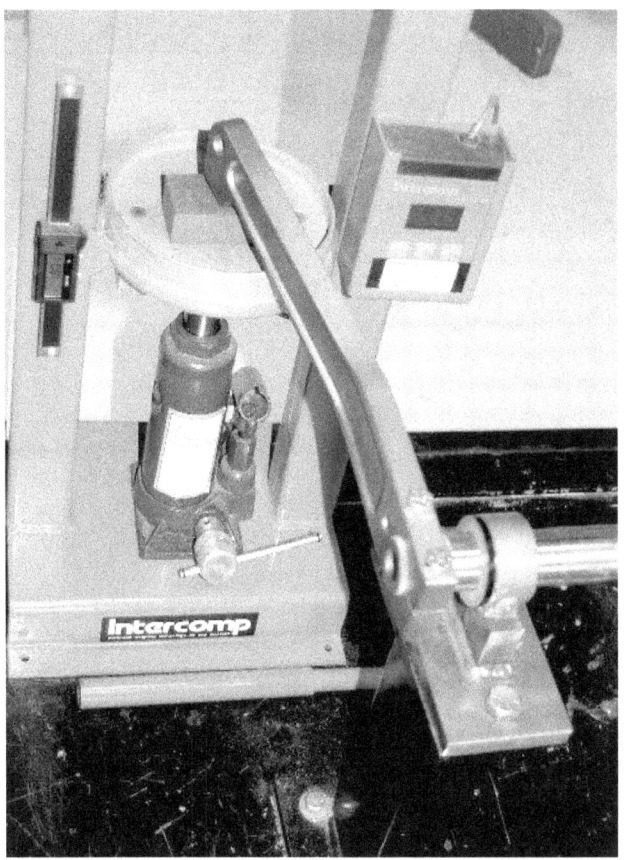

*In years past, we may have rated the sway bar using a scale pad or spring rate like in this illustration. This tells us exactly what the rate of the bar is at the end of the arm in pounds per inch. But it does not tell us how much work it is doing in combination with the springs at the motion ratio our springs perform at.*

### Shear Modulus Numbers for Types of Metals

| Metal | Value |
|---|---|
| Cold Rolled Steel | $11.5 \times 10^6$ |
| Nickel Steel | $11.0 \times 10^6$ |
| Iconel (nickel-chrome-iron alloy) | $11.5 \times 10^6$ |
| Molybdenum Steel Compound | $17.1 \times 10^6$ |
| Almost all ultra-high strength steels contain molybdenum in amounts from 0.25 to 8%. | |
| Stainless Steel | $10.5 \times 10^6$ |
| Carbon Steel | $11.5 \times 10^6$ |
| Titanium | $6.5 \times 10^6$ |

*This is the chart showing different Modulus numbers for different types of steel. We can easily see where a formula might not give us the correct information due to the many different modulus numbers and the fact we don't know exactly which metal our sway bars are made of.*

## Formula for Determining Sway Bar Rate

$$\text{Rate} = \frac{SM \times (D^4 - d^4) \times Pi}{L \times A^2 \times 32}$$

Where:
SM = Shear Modulus
D = Diameter of the Sway Bar in inches and decimals
d = Inside diameter of a hollow sway bar
L = Effective Length of the sway bar in inches and decimals
A = Effective Arm length in inches and decimals
Pi = 3.1416
Rate = Bar rate in pounds per inch of arm movement

*Here is the formula for finding the rate of a sway bar or torsion bar and that rate is the rate at the end of the arm. Note again that the Shear Modulus is the number that is hard to be sure of, and that makes using a formula problematic.*

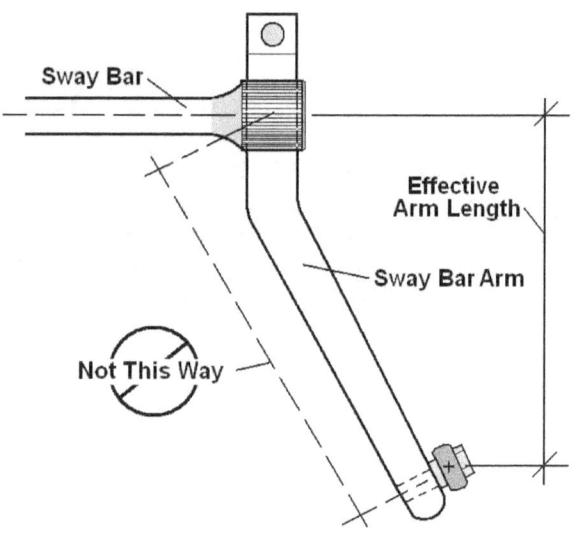

*If you decide to calculate your sway bar rate, you need to measure the arm lengths at a distance that is at ninety degrees off of the sway bar, not at an angle. This sketch shows the correct way to measure.*

**Conclusion** – What you find for rates for your sway bars might be what you expected or not. This is why we go through this type of exercise. We need to know every influence that affects our setups. If our sway bar force exceeds half the overall force we need at mid-turn, then maybe we need a softer rate of sway bar.

In any case, you now know how hard your sway bar is working and you can make some decisions about what size bar to use. Having this type of information is the key to being able to properly plan out your overall setup.

# Exam - In The Context Of This Lesson:

### A Sway Bar or Anti-Roll Bar Does What?

1) Keep the car from rolling
2) Helps to roll the car
3) Resists chassis roll
4) Keeps the chassis balanced

### What Affects The Rate Of A Sway Bar?

1) The diameter of the bar
2) The length of the arms
3) The length of the bar
4) The modulus (hardness) of the metal
5) All of the above

### We Can Accurately Rate A Sway Bar By What Method?

1) Using a formula
2) Using a chart from the manufacturer
3) Making the sway bar support a known weight
4) Measuring the sway bar force as wheel rate or spring rate
5) 3 and 4

### A Sway Bar Assist The Ride Springs By?

1) Adding spring rate to the suspension
2) Providing more wheel rate
3) Providing roll stiffness support
4) All of the above

### To Determine The Correct Mid-turn Ride Spring Force, We?

1) Add the sway bar force to the wheel rate
2) Add the sway bar force to the mid-turn force
3) Subtract the sway bar force from the mid-turn force
4) Add the sway bar rate to the wheel rate

## Lesson Fifteen – Shock Selection Matching Spring Rates

In previous Lessons we have learned what shocks do and cannot do. That is important to understand if we are going to be successful with developing a shock program for a race car. Some race teams continually try to make the car do things with their shocks that could be better served with other settings and components.

Here we will be setting up a race car with shocks based on what each shock will be working with, and those are forces. All components that contribute to the generation of force in a race car suspension must be controlled by the shocks while the chassis and suspension is in motion. In motion. To re-state, a shock does nothing if it is not in motion.

*On exit, we have weight transfer and some suspension movement. In this case, with the suspension movement, shocks can help us distribute the loading on the four tires differently to enhance our transitional handling.*

*We experimented with Dick Andersons super late model car some years ago to find the proper front shock rates to run with the bump springs we installed. We matched the shocks to the spring rates on the front and back of the car and had a very fast machine. A lot has changed in the layout of both the spring rates and the shocks since then. What we present in this Lesson is inline with the current thinking. The shocks on this car at mid-turn is not influenced by the rates of the shocks because they are not in motion. This is called a steady-state condition because neither suspension is moving.*

If we expect the shocks to transform or simply control our suspension, then we can only expect the shocks to contribute when the suspension is in motion. We need to think out time when the chassis moves, how it moves and at what part of the track it is moving.

The other major consideration is: how does the control of movement of one shock affect the movement and control of the other three shocks on the car? It is this interrelationship of the four shocks and springs that will ultimately determine where the loads on the tires go during the transition periods, or those times when the suspension is moving.

Based on the above statements, we can now declare that **shocks control the distribution of loading on the four tires while the chassis is moving**. Movement can be defined as: 1) chassis dive on corner entry from braking and weight transfer from the rear to the front, 2) chassis squat on corner exit under acceleration from weight transfer from the front to the rear, and 3) chassis roll throughout the corner from lateral centrifugal force that rolls the chassis and transfers weight from the inside tires to the outside tires.

**Working With Shocks At The Track** – Everything we do to tune the car with shocks should be based on the knowledge we have learned above. We will get into specifics about each corner of the car and each of the transition segments of the race track. We will study the motions for each transition and how those motions can be used to enhance our setup and improve the performance of the race car through the segments.

There are many different types of setup these days and racers are trying to do things that the laws of physics just won't allow. You don't have to be an engineer to understand the problems that can come as a result of this. With just a basic understanding of how the dynamics of the race car work, we can all choose our shock rates correctly to complement our setups.

In the modern world of short track racing for both dirt and asphalt competition, shocks have become one of the most important tuning tools we have. They can complement the current setups, especially the radical bump setups if applied correctly.

The following information is useful whether for teams who are running either the more conventional setups or the more radical bump setup ones. Shocks affect the speed of movement and the load distribution at the four corners of the car as we transition from high speed to minimum mid-turn, and then back up too high speed again.

One of the most basic and important things to remember is this. Shocks do not affect the handling of the car if they are not in motion. To say that you can tune your mid-turn, steady state handling with shock changes is wrong. Any competent shock expert will tell you that.

**Shocks Work With Springs** - Controlling wheel movement would be much easier if all we had to work with was the shocks. But in reality, our race cars are supported by a set of springs and other components that produce forces. Basically, we always want to match our shock rates to the suspension forces of: 1) the ride springs, 2) the sway bar, and 3) the bump devise you might be using. In all situations, the shocks rebound rate will always need to be greater than the compression rate for any racing shock because the springs help resist compression and promotes rebound.

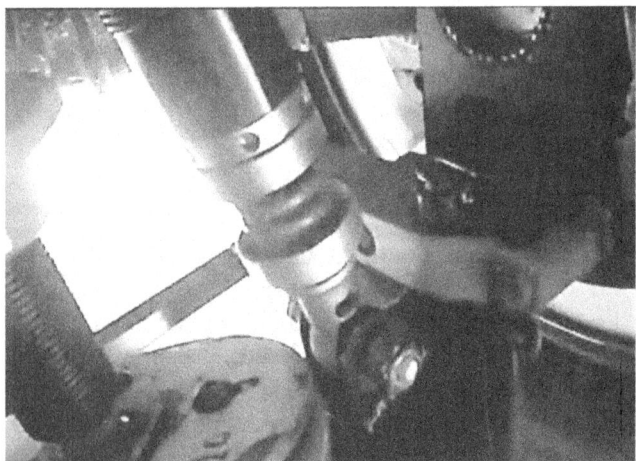

*This image taken from a video shows an early version of the bump spring in action. The shock rates need to be high enough in rebound control to manage this high spring rate, in this case 1200 ppi.*

https://www.youtube.com/watch?v=u9clWQpps6Q

*This video shows a late model circle track car on-track with a 1200 ppi RF bump spring installed. This is a fairly bumpy track and we can see that the motion is enough to almost coil bind the spring. We could have used either a stiffer rebound shock or a stiffer bump spring.*

https://www.youtube.com/watch?v=RlvO7tDJm4Q

*In this video, we ran the same car with a stiffer 1500 ppi bump spring and the motion was reduced. Modern bump springs are available in much higher spring rates that would further limit the motion of the suspension on bumpy tracks.*

As we install stiffer springs, we would naturally need to increase the rebound resistance and decrease the compression resistance. Stiffer springs would include adding bumps to one or more corners of the car. Each bump device has a spring rate in the range of motion it is operating in. Bump springs have a constant rate that is just added to the equivalent ride spring rate.

Let's expand on this thought. A stiffer spring that supports the same load, or generates the same force, will rebound faster than a softer spring supporting the same load. So, even though you are not generating more force to carry the intended load, your rebound speed will increase with the spring that has a stiffer spring rate.

**Conventional Vs. Bump Forces** – The actual force needed to support the car at all points on the race track is no different between the conventional setups and ones what use bump stops or bump springs. The natural forces of gravity, centrifugal (lateral) force and mechanical downforce due to track banking are the very same if the turn speeds remain the same.

The difference between the setups is the dynamic spring rates, or stiffness of spring the suspension is using during the trip through the turns. A conventional setup using 250 ppi (pounds per inch) coil-over springs will rebound with less force than a bump setup running on 1500 ppi bump springs or bump stops that rate in that range.

The shock rates we need to use to control the conventional setup is much different and softer than the ones we will use to control the much higher bump setups spring rates. It is all about controlling the speed of movement during times of load transfer. Obviously a stiffer spring will rebound faster when X amount of weight is put on it, or taken off of it than a softer spring with the same X amount of weight change.

**Soft Spring Setups Are Not** – In a soft spring setup running on bumps, the setup is not really soft. The ride springs are soft to allow the car to be pushed down onto the bump devices and then the suspension becomes very stiff with a very high spring rate.

So, we may say we have a soft spring setup because we install 150 ppi ride springs at the front in our coil-over circle track race car, but when we add the stiff bumps to that, we end up with 1200-1500 ppi or more spring rates. That is why I avoid saying, soft spring setups when referring to bumps setups.

Therefore, with those setups, keeping with the idea that the shocks must control the installed spring rate, we must run shocks with a rebound rate upwards of 1200-1500 pounds at 3-5 inches per second of speed.

**"Tie-Down" Shock Terminology** – This brings us to an important discussion. The use of the term "tie-down" has been around for some time to describe shocks that are high in rebound resistance. The idea initially was that if we install these high rebound shocks, we can tie that corner of the car down and keep the tire more in contact with the track surface. This way of thinking is completely wrong.

First off, a tire is not attached to the track surface, it is free to rise and fall according to what the other three corners of the car are doing at that instant on the track. If our setup wants that corner to move vertically because load has shifted off of it, then the load will come off regardless of how much we "tie down" with shocks. You just won't see much suspension movement.

So, we might be fooled into thinking that it indeed did work. No, it did not. A suspension can lose or gain loading without moving. Movement in and of itself is no indication of tire loading. If enough load comes off that corner, or the setup is unbalanced enough, the tire might well come off the track surface even without suspension movement.

The unloading of the tire will occur on the inside front or inside rear corners in most cases. So, we cannot tie the left front tire down and we cannot tie the left rear corner down, although I've seen some teams try. Lack of movement of the suspension does not mean that load remains on the tire.

**Entry Tuning** – If we split the front shock compression rates with a RF shock using a stiffer compression rate than the LF shock, then, while the suspension is in motion due to load being transferred to the front on entry, the RF suspension will move slower than the LF suspension. Additional load will be transferred onto the RF and LR tires causing a momentary increase in the corresponding diagonal weight percentage in the car.

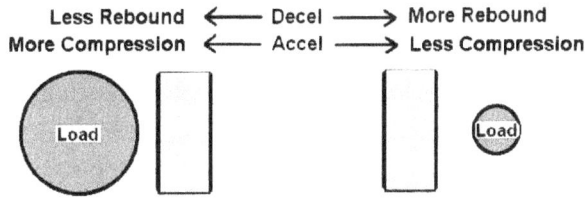

*In this example, for the rear tires, the larger circle represents more tire loading. Under deceleration when the car is braking and weight is transferred from the rear to the front, less rebound in the LR shock leaves more retained load on the LR tire and conversely, more rebound in the RR shock removes more weight from that tire. Under acceleration, more compression resistance in the LR shock causes the LR tire to support more of the transferred weight over the RR tire and conversely, less compression in the RR shock causes less load to be transferred onto that tire.*

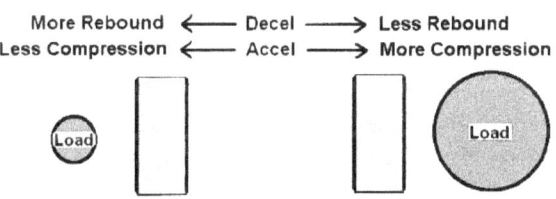

*This shows the opposite results for rear tires while under acceleration and deceleration. On Decel, more rebound in the LR shock pulls more load off that tire and less rebound causes the RR to retain more loading on Decel. On Acceleration, more compression in the RR shock will load that corner with more of the weight that is being transferred.*

The affect is more pronounced with conventional setups that might be used now days in the classes that are restricted from using bumps, or the stock classes. That is because those suspensions move much more than ones running on bumps. But it can still affect the bump setup cars to a certain degree.

It is important to note here that during the transitions, the load transfers almost immediately when a force is presented to enact that transfer such as applying the brakes. As we apply brakes entering the corner, the load transfer to the front happens quickly. If on entry we transfer 300 pounds from the rear to the front, the 300 pounds goes to the front in an instant.

The distribution of that 300 pounds between the two front tires, while the suspension is in motion and assuming a new attitude, will depend entirely on differences in stiffness of the suspension systems at all four corners. Stiffness is defined as the resistance to movement influenced by the shocks and springs.

The result of all of the above is this. The slower moving (or stiffer) corner will momentarily retain more of the transferred load while the suspension is in motion traveling to a new attitude. This added loading also affects the diagonal corner and loads it as well.

Cross weight in circle track racing is defined as the percentage derived from the combined RF and LR weight divided by the total vehicle weight. If the cross weight percent increases, then the car will be tighter on entry. This is exactly why it has been said that a stiffer RF shock will speed up load transfer to that corner, although we won't be using that terminology here.

If we install a LF shock that is stiffer than the RF shock, and/or run a stiffer LF spring than on the RF, then we can effectively reduce the cross weight in the car on entry while the suspension is in transition by loading the opposite diagonal, the LF and RR. As one diagonal goes up in percentage of supported weight the other goes down.

Modern bump setups use much lower front compression settings due to the high spring rate of the bumps. We can still utilize this method by creating a compression split. The effect will be less than with a conventional setup, but still somewhat effective.

Once the car has decelerated enough to make the turn, and the shocks are firmly on the bumps, the load distribution will equalize. Therefore, the advantage of shock compression split will be nullified. If the shock rebound settings are high enough, the shock will never leave the bump and no amount of compression split will work because there will be very little shock movement.

We can see where with conventional setups, we can probably utilize differences in shock compression rates where with the bump setups, the restricted movement nullifies the effect. We are better served working with the rear shocks only for the bumps type of setups.

**Exit Tuning Using Split Valve Shocks** – Corner exit performance can be improved by utilizing the shocks. This is done by either splitting the compression settings in the rear shocks and/or utilizing the rebound settings in the front shocks. In a car with equal rear spring rates, a stiffer compression setting in the LR shock than in the RR shock will load the LR and RF corners as weight transfers onto the springs.

Load is transferred to the rear under acceleration and while the rear suspension is in motion, this split will tighten the car by increasing the cross weight percent. But, this event is very short lived. Because as we stated, weight transfers very quickly, so the effect of shocks happens just as quickly and goes away just as quickly.

For all setups, a shock with a stiffer rebound rate at the LF corner can help accomplish the same effect by causing a slower movement of that suspension and a more rapid transfer of load off of that corner and the diagonal RR corner, which will in-turn increase the percentage of load supported by the RF and LR tires.

With the bump setups, the rear compression settings can really help a car that is otherwise limited in adjustments. If the RR spring is somewhat stiffer than the LR spring, and this is very common with bump setups, there will be a loosening effect on acceleration when load transfers to the rear. The stiffer RR spring causes load to be placed on the RR and LF corners reducing the cross weight percent.

By installing a LR shock with a stiffer compression rate, and a RR shock with a softer compression rate, you can nullify the negative effect of the stiffer RR spring. This serves to equalize the unequal resistance to compression due to the dissimilar spring rates and helps keep the car tight on exit. This again is a short lived event.

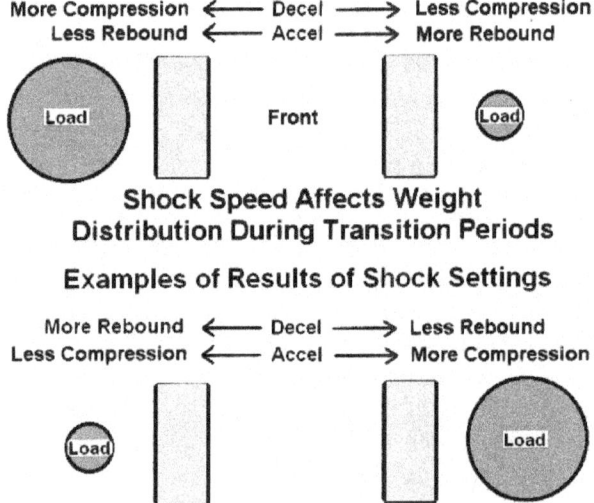

*Here we can see the results of shock rate split for both the front and rear shocks and how the corners react. To load the RF and RR tires for either Accel or Decel, we follow along with the suggestions.*

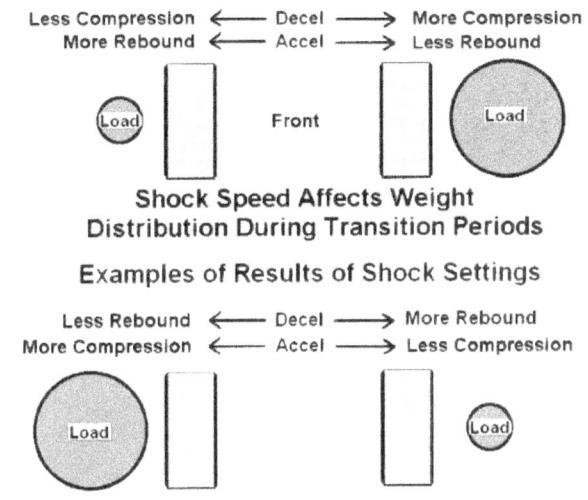

*To load the RF and LR tires for Accel or Decel, we follow the suggestions here to see the entire picture of how the four tires are affected by shock rate differences side to side.*

**Putting All Of This to Use** – In order to utilize the configurations we have discussed above, we must be able to use a range of different rates of shocks in order to find the right combination for a race car at a particular race track for a particular setup. For a team that races at only one track, the process is fairly simple.

You would need to experiment to find the fastest set of shocks and ones that suit the driver's style and stick to those staying within the boundaries of physics. For teams that travel to different tracks, some changes might be necessary if the setup (read as spring rates) needs to change and/or the track layout is different from track to track.

Most shock experts agree with the following basics:

1) The shock package should be softer overall when racing on dirt and when the track is flatter when on asphalt for the conventional setups.

2) Get your basic setup close to being balanced before trying to tune with shocks. Shocks cannot solve basic handling balance problems that occur at mid-turn when the shocks are not in motion.

3) Higher banked tracks require a higher overall rate of shocks and springs as opposed to flat tracks. This is because of the higher speeds and the extreme amount of suspension forces.

4) Shocks that are mounted farther from the ball joint should be stiffer than if they were mounted close to the ball joint. That is because with each inch of travel of the wheel, the shock mounted farther away will move at a slower speed which means less resistance in both rebound and compression. This is also true of shocks mounted at high angles to the direction of motion.

5) Of the two transitions, tune entry performance first with your shocks. If there are no entry problems, make small changes if you want to experiment to see if entry speeds can be improved. Entry problems include a tight or a loose car.

By far the worst problem would be the loose-in condition. This can involve an alignment problem, but far too many times it is as we have discussed, a inside rear shock that is too stiff in rebound. For road racing cars, this could include both rear shocks.

6) Tune exit performance last. If there are no exit problems, don't make any significant changes. Exit problems can include a car that pushes (under steers) under acceleration or one that goes loose (over steers) under power. Be sure that you do not have a Tight / Loose condition where the car is basically tight in the middle and goes loose just past mid-turn. This is fixed with spring rate and/or moment center adjustments, etc.

7) On dirt race tracks, reduce rebound settings on the left side and decrease the compression rates on the right side for dry slick surfaces to promote more chassis movement. This helps to maintain grip as the car goes through the transitional phases of entry and exit.

8) For the bump setups on asphalt, the whole shock package must be much different than when running conventional or soft conventional setups. The bump spring rates (either bump stops, or bump springs) will be very high and so the shock rebound rates must match those high rates in order to control the ride spring and bump device.

Using a bump that is rated in the 1000 pound per inch spring rate range will need a shock that is rated at around 1000 ppi at 3 - 5 inches of movement per second. Usually the low speed rate of the shock will be comparatively high too and we often see a "nose" rate of between 500 and 800 ppi or more at less than one inch per second speed of movement.

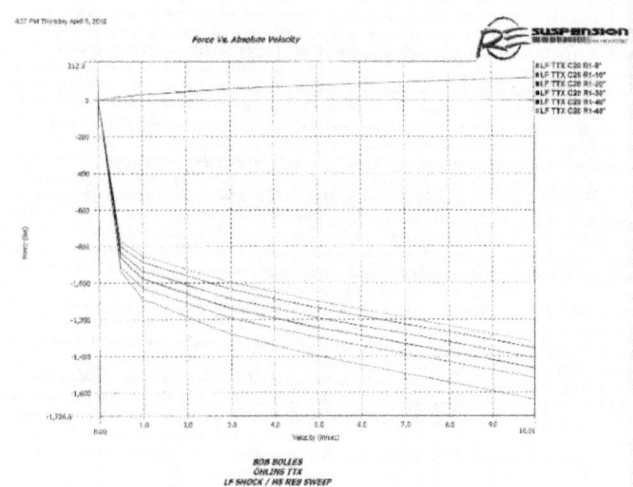

*These are typical shock rebound and compression rates for a Left Front shock when running on very stiff bump setups. Note that at 3.0 inches per second of Velocity, we are in the range of adjustment of between 1,000 pounds up to about 1280 pounds. The rate builds quickly from motionless up to 0.5 inches per second of speed. It then builds more slowly up to 10.0 inches per second of velocity. Each line represents the force of the shock at an adjustment of the shock bleed valve. The bottom line on the rebound side represents a closed bleed valve. The other five lines represent openings of the bleed valve from 10 "clicks" to 48 "clicks".*

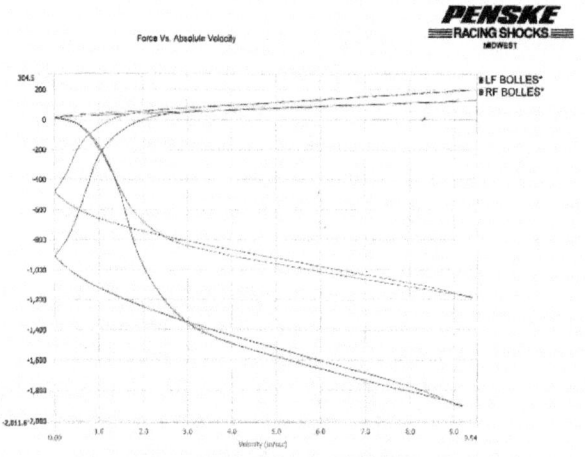

*In advanced tuning for both the LF and RF shocks on a car running bumps setups, there is very little compression (lines above the zero force mark) as the bump provides plenty of compression resistance, so the shock doesn't have to do much work in compression. In rebound (lines below the zero force mark) we have what is called a "nose" rate. The red line, or RF shock, has about 500 pounds of force resistance to movement before the shock even starts to move. Then the force builds up to about 700+ pounds of force at 3.0 ips of velocity. The LF shock has about 900 pounds of "nose" force at zero movement and over 1,300 pounds of force at 3.0 ips velocity. We generally control the LF spring and bump to their rates and control the RF bump to about half of their rates for circle track cars turning left.*

**How To Make Changes** – Racing shocks are either adjustable or not. The adjustable shocks might have rebound adjustments only, or both rebound and compression adjustments. For non-adjustable shocks, you would then need to physically change the valving in the shock, or have someone else do it. If the shocks are not re-valvable, then you would need to acquire shocks with a different value.

Every race team needs to have a shock dynamometer or send their shocks to someone who does, to be rated. The team must know the force of resistance of each shock on the car represented by a shock graph generated by a shock dyno. If you do not have this information, then you are guessing at the amount of work and the forces your shocks produce. No part on a race car can afford to be guessed at.

**Summary** – The suggestions provided here are representative of trends that can enhance your handling package. Before any of this can work, the setup must be balanced, the steering characteristics must be ideal and the car must be aligned properly. If not, you will probably chase the setup and experience a lot of frustration and expense. Shocks are to be used as a tuning device only after all of the other settings on the car have been perfected.

Shock tuning is the last thing to experiment with in order to try to increase your race cars performance, but it is nonetheless a necessary step in finding the ideal total handling package. That said, before you setup your car and chose your shocks, evaluate what you will need as far as shock rates that will match the spring and bump rates you will run.

When we experiment with the bump setups, we would do well to consult with the shock experts so that we can match our shocks to the spring rates and overall forces we will be working with. Most racing shock companies have technicians who are very familiar with those setups and can advise as to the best rates to use to match those bumps.

Because there are so many different setups, we cannot suggest specific rates of shock to use. This Lesson is intended to help you think out your changes as you evaluate your setup and try to make improvements. In RCT Level Three, we will be working with actual race cars and developing shock packages to help those setups.

# Exam - In The Context Of This Lesson:

## Shocks Control What?

1) Speed of suspension movement
2) Suspension forces
3) The rate of the spring
4) Transitional handling
5) All of the above

## A Higher Rebound LR Shock Than The RR Shock Does What?

1) Loads the LF and RR corners on acceleration
2) Loads the LF and RR corners on deceleration
3) Loads the RF and LR corners on acceleration
4) Loads the RF and LR corners on deceleration

## A Higher Rebound LF Shock Than The RF Shock Does What?

1) Loads the LF and RR corners on acceleration
2) Loads the LF and RR corners on deceleration
3) Loads the RF and LR corners on acceleration
4) Loads the RF and LR corners on deceleration

## A Higher Compression LR Shock Than The RR Shock Does What?

1) Loads the LF and RR corners on acceleration
2) Loads the LF and RR corners on deceleration
3) Loads the RF and LR corners on acceleration
4) Loads the RF and LR corners on deceleration

## A Higher Compression LF Shock Than The RF Shock Does What?

1) Loads the LF and RR corners on acceleration
2) Loads the LF and RR corners on deceleration
3) Loads the RF and LR corners on acceleration
4) Loads the RF and LR corners on deceleration

## To Control A Stiff Spring, We Need To?

1) Have more compression control and less rebound
2) Have more rebound control and less compression
3) Make the rebound and compression the same
4) Run more compression than rebound control

## To Control A 150 ppi Ride Spring Plus A 1350 ppi Bump Spring, We Need What Shock Rebound Rate?

1) 1500 pounds of force at 10.0 ips velocity
2) A nose force of 1350 pounds
3) 150 pounds of force at 3.0 ips velocity
4) 1500 pounds of force at 3.0 ips velocity

## Lesson Sixteen – Bump Setups Cause and Effect

Race car suspensions and setups have gone through many changes in recent years. Ever since the onset of the "Big Bar Soft Spring" era, racers have found themselves trying to go softer and softer with their suspensions to achieve faster turn speeds. The trend started in Cup racing when those teams started running coil-bind setups intended to lower the race car on the track to improve the aero effects.

Now that the short track industry and even the road racing community including prototypes and formula cars have now started to use bumps in their setups, it is time to explain how to utilize bump technology to go faster.

*The stock car industry has taken to setups utilizing bump devices such as bump stops and bump springs. The reasons why include a lower front end to produce a lower center of gravity and better aerodynamic properties, and a consistent ride height which reduces camber change, something tires like. To get this attitude, the ride springs need to be soft, the bumps need to be stiff and the shocks need to be of a sufficient rebound rate to control the stiff bumps.*

Every expert will agree that bump setups offer the following three distinct advantages over conventional setups. They are:

1) *Lower center of gravity.* This is important because the weight that is lower is at the front in stock and circle track cars, and the engine is a big part of the weight up front. Lower that chunk of metal and you've really lowered the CG by quite a bit. A lower CG transfers less weight in the turns and more weight ends up on the left, or inside, tire providing more equally loaded front tires and more grip.

2) *Better aerodynamics.* A lower front end and valance cuts the air that would go under the car and provides more low-pressure area under the front bodywork for more downforce. Also, it provides more angle of the hood area above the engine and this increases the effect of flat plate aero and offers more downforce.

3) *Reduced Camber Change.* Finally, a race car on bumps moves vertically very little and therefore there is very little camber change in the tires. Race car tires don't like camber change. The elimination of camber change could be the most important advantage in using bump setups.

Now in our current situation in racing, bumps are being used on almost every corner of the car. Certain advantages can be had by using a bump device for increasing the spring rate and/or loading on one or more corners of the race car.

There are several types of bump devices used in racing today. The first ones on the market were loosely called bump stops because they stopped the soft spring suspensions from moving too far and were used to prevent the car from bottoming out on the track. Even road racing prototypes used bump stops on high banked and fast tracks like Daytona.

A variation on the bump stops were the bevel washers that provided a consistent spring rate, but the available travel was not much more than the bump stops. Then the bump springs hit the market and a whole new and exciting way to bump became available.

The bump stops were hard on suspensions when run on rough tracks, and the bevel washers bottomed out. The bump springs offered considerably more travel and a consistent spring rate. Here are detailed explanations and applications of each.

*This is one type of bump stop made out of fairly soft foam material. These are available in stiffer materials for different rates and properties. They can be combined with other bump stop designs. These are installed on the shock shaft and are compressed between the bottom of the shock body and a retainer cup at the bottom of the shock shaft.*

**Bump Stops** - Bump Stops are one of the best methods of achieving the three bump advantages. When using bump stops, you install a stack of properly selected bump stops on one or both front shock shafts. You then select a front spring and sway bar package that will give you the best on track body attitude. Bump stops work best when used on a smooth track surface.

The key is knowing and achieving the proper wheel load in critical parts of the corner to balance the chassis handling. We do this by selecting different hardness ratings in the bump stops themselves, and by adding spacers or packers to alter the timing and engagement points of the shock body to the bump stop stack.

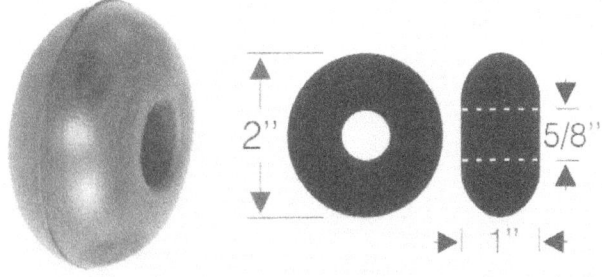

*This is another type of bump stop and is a stiffer spring rate and can be stacked to provide a softer rate over using just one.*

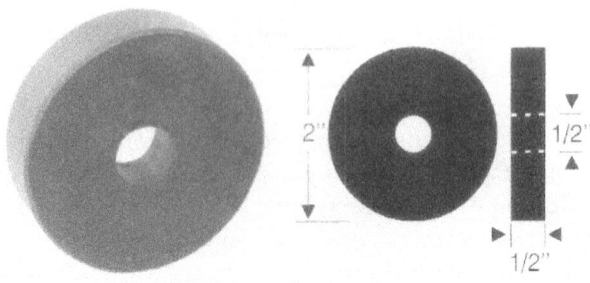

*Here is yet another type of bump stop referred to as a "hockey puck" due to its resemblance to that. It too can be stacked and is available in different densities for a different spring rate. All bump stops are rising rate springs. That is, the spring rate rises as the bump stop is compressed.*

This in effect creates a specific amount of wheel load for each wheel, based on a given amount of suspension travel to achieve that optimum body attitude and optimum distribution of wheel loads at the same time. We have force measurement rigs to apply the shock travel to the coil-over with the bump stops and then read the force it is under.

Here is an example: A traditional balanced setup with a 300 ppi RF ride spring, produces 900 lbs of force at 3" of shock travel to support the weight of the RF corner of the car at ride height. If this spring traveled an additional 3 ½ inches, the spring force would equal 1950 lbs. The problem is getting the spring to compress early enough to provide the benefits that staying on the bumps provides.

Alternately a balanced bump stop set up might utilize a 150 ppi RF Spring that is preloaded to provide the same 900 lbs of force needed at ride height. Then when the braking and corner loads are added to that spring, it easily moves down because of its soft rate and is stopped by the bump stop. Once it engages the bump stop stack at the same 3 ½" of shock travel from ride height, the force goes up immediately to the required 1800 lbs or more of total force to hold the car up against the turning forces. As you can see both setups are now achieving the about same wheel load at max travel keeping the balance very similar, but one very important difference exists.

The bump stop setup with softer springs reaches its optimum load sooner, and maintains it for a much longer time frame through the corner. With the right shocks installed, the shock will stay on the bump stop most of the way around the track. This has several major advantages.

First, as we discussed earlier, the camber will be at its optimum settings for maximum grip much earlier in corner entry, through center and remain longer through corner exit. The overall center of gravity of the car will also remain lower throughout the entire lap, also enhancing every part of the corner for more overall speed. Lastly, by using softer springs in the front, the front end will stay down longer, and will therefore offer better aero downforce throughout the turns from initial turn-in to corner exit.

The simplicity of a good bump system and ease of fine adjustments make bump stops the number one choice for a suspension package wherever rules allow. There are tools available now that help you determine the forces your bump package produces and then you can fine tune the loads on the bumps to maximize the loading on the four tires.

Drivers first impressions of a well-balanced bump setup are usually that they have much more feel in the front of the car. The cars usually respond very quickly to small inputs in steering, and less overall steering input is usually required.

*The very beginning of the trend to create a lower front ride height came from using coil bind where when the softer ride springs compressed, then ran out of travel and then the springs coils contacted each other. This prevented the car from bottoming out on the race track, but the suspension now had no spring rate except for the tires spring rate. The suspension was basically solid at this point. There had to be a better way, and bump stops was the next step.*

Initial experimentation with softer springs involved letting the spring go into what we call coil bind. The spring ran out of travel and the coils contacted each other and that was what kept the car from bottoming out. It was a harsh way to go and did provide the benefits, but the ride was harsh to say the least.

Bump stops have an advantage over coil-binding because they do compress somewhat and the ride improved when teams started using bump stops. With bump stops you can fine tune the weight distribution and cross weight settings much more easily without changing any critical dynamic chassis heights or stop heights. Bump stops can also provide much more compliance than the coil-bind setup making it a better choice when rough track surfaces are a concern.

*When we use bump stops, we need to know the shock travel for mid-turn so we can duplicate that travel on a spring force rig. Once we know the force, we can adjust the force to what we need, no more and no less.*

A well-balanced bump stop setup using soft springs is much closer in relative wheel rate to our older traditional setups than many racers realize. The key word used here is balanced. We talked about balance before and the need to keep the two ends of the car working together. With the bump setups, this is still the primary goal and balance provides the platform for proper load distribution through the turns.

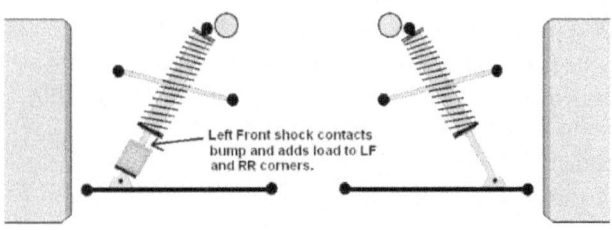

*For many asphalt circle track race teams, only one bump is used on the car and it is mostly run on the LF corner. This forces a lot of loading onto the LF tire and may overload that tire, rather than providing a balance of loading on the two front tires.*

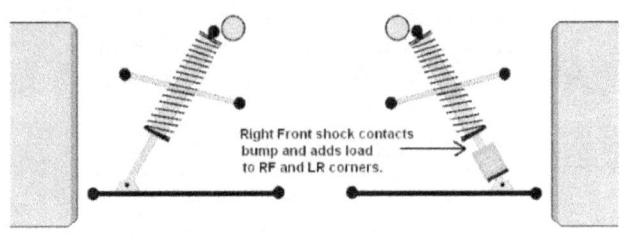

*For dirt circle track late models, the bump is often used on the RF corner. This loads the RF and LR under braking and cornering. For slick tracks, it is an advantage to load two tires heavily in order to cut through the slick layer of sand to get down to the hard clay and to get more grip.*

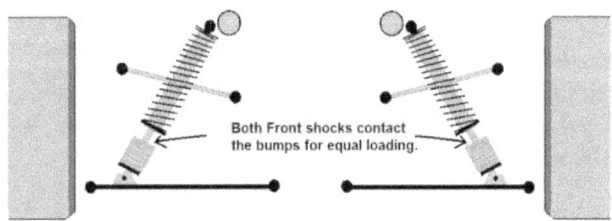

*A better way to setup an asphalt circle track race car, or a road racing stock car, is to install bump stops or bump springs on both the front shocks. This balances out the forces and the setup providing better load distribution and a more balanced setup. The method we talked about for setting the cross weight on the bumps cannot be done when you only have one corner on the bumps. We can then see where there will be a drastic change in the weight distribution when the chassis transitions from ride height off the bump onto one bump.*

**Bump Springs** - Things are changing fast in the bump arena. A couple of years ago, someone introduced a commercial bump spring. The concept of bump springs was not new and racers have been experimenting for many years with using various springs such as valve springs to use as additions to the softer ride springs. But the bump spring had never been commercially available.

*This is the first commercially available bump spring produced and sold by Chassis R&D. It was an adaptation of a flat coil spring called a "die spring" that was normally used in industrial die stamping machines. It proved that using a spring with a consistent spring rate could be a viable alternative to bump stops. It provided more travel than a bump stop, a consistent spring rate throughout its range of travel, and it was durable unlike many bump stops that deteriorate quickly. The available travel for this spring is 0.75 inches.*

Now we see where most major racing spring companies offer their version of a bump springs. Here we will explain what a bump spring is, what it does and how it may be superior in some cases to other bump devices.

Traditional bump stops are non-linear in their spring rate. This means that the farther they are compressed, the more spring rate they develop. This tends to make your setup tricky to balance and sometimes unpredictable. The traditional bumps also have very little travel available to absorb bumps and holes in a rough race track.

With the Bump Spring, we have a consistent spring rate at all positions. We also have much more travel available to smooth out the rough tracks we sometimes run on. The traditional bump stops usually allow around a quarter inch of movement and go to a very high spring rate approaching maximum travel. Most bump springs have three-quarters to one full inch of travel before coil bind and the spring rate never changes.

*Bump springs are available in many different spring rates and heights. This collection of the first commercial bump spring was available in rates of 1,000, 1,200 and 1,500 ppi rates. Now, all major racing spring companies offer their version of the bump spring in many different rates and configurations.*

So, we can see where the bump spring has characteristics that are better than the other bump devices due to the increased available travel and the consistency in spring rate. The fact that most of the racing spring companies have developed their own line of bump springs speaks to the fact that they may be better for many applications for teams choosing bump setups.

In reality, by using bump springs, we are actually going back to yesteryear and the stiff spring setups from the 1970's to the 1990's. The only change is that now we can run the car much lower and take advantage of the improved aero, a lower center of gravity and the reduction of camber change.

**Setting Cross Weight For Bump Setups** – Once the shocks are firmly on the bump springs, the load distribution will change very little since there is little movement of the shock through the turns. So, if you have not pre-set the cross weight percent with the car on the bumps, it may change dramatically from what was set at ride height when the car is off of the bump springs.

The following is how you should set your weight distribution when running bump setups. Failure to do this can result in a significant change in weight distribution from when the car is at ride height to when it contacts the bumps.

*This shows how bottom of the shock comes in contact with the bumps spring. This shock would need a screw mount for its top mount in order to adjust its spacing, or the timing when the shock contacts the spring on entry and through the middle of the turns. Other applications use spacers, or what are called packers, to adjust the point in shock travel where the shock body contacts the spring.*

You can set your cross weight both on and off the bump stops or springs. The procedure is as follows:

1) Begin by setting the correct cross weight percent on the scales at normal ride height;

2) Now remove the ride springs and allow the car to rest on the bump springs or bump stops with the bumps supporting the weight of the front end;

3) Bring the corner weights back to what you had at normal ride height by adjusting only the amount of spacing, or packers, between the front shocks and the bump springs or bump stops. If you have height adjustable shock mounts, just make changes to the shock heights to get your static cross weight number. Then just reinstall your ride springs and go racing.

*Today's bump spring is of a round coil bar design just like the ride springs. They are designed to be compact and have significant travel for a bump device. And they are offered in a wide range of rates for any application and loading required.*

This way, your car will not experience a change in load distribution when transitioning from ride springs onto the bump springs. This seems to be a very common problem when using any type of bumps.

**Load Changes Due To Sway Bar Loading** – Another effect caused by the bump setups is the loading of the sway bar when the front travels to the bumps. On many cars, if you were to set your sway bar to neutral at ride height, when the car is down on the bump springs, the sway bar will load by several turns because the two sway bar arms don't move the same distance. The right sway bar arm travels more than the left sway bar arm and that loads, or twists, the bar.

This definitely causes a change in the load distribution at the four wheels. When using larger sway bars, the change is much greater. What you can do is set your bar to neutral at normal ride height and then put a few crew members on the front of the car and push it down onto the bumps. While on the bumps, check to see how many turns it takes to return the bar to a neutral setting with no pre-loaded.

Once you know how many turns are involved, just set to neutral at ride height and then back off that number of turns. Or, if you are using a smaller sway bar, say 1.25 inches or less, you can just leave it preloaded when you set your cross weight on the bumps as we described above. That way, you can run pre-load on the sway bar for bite off the corner while still maintaining your correct cross weight percentage.

Another important thing to remember about the sway bar is that it provides added spring rate to the suspension. It acts, in roll, like the car has a stiffer left and right side ride spring. When the car rolls, the sway bar resists the rolling motion and uses its spring rate to do that. So, if we know what force we need in the suspension to support the car at mid-turn, we need to take into consideration the sway bar force that is working in addition to the spring and bump forces. If we don't consider the sway bar force, we may overload that corner of the car with too much ride spring force.

**Rear Spring Rates For Bump Setups** – Initially when teams started running front end bump setups, the only way to match the front and rear roll angles, and balance the setup, was to run a spring split in the rear with the RR spring much stiffer than the LR spring. Many teams overdid this split and tried to run much more split than was necessary.

The problem with running this spring split is that on corner exit, the weight transfers onto the rear springs and the LR spring compresses more than the RR spring. This loads the LF and RR corners and loosens the setup by reducing the cross weight percent.

To fix this, the teams noted the shock travels using the stiffer RR spring and then recorded the force that amount of travel produced. Then they installed a much softer RR spring that was preloaded some amount to help provide the force needed at mid-turn to support the car and provide the balance needed.

Then when the car accelerated, and weight transferred to the rear springs, both rear springs compressed about the same amount and the weight distribution remained unchanged. Some teams are even experimenting with a RR spring rate that is actually softer than the LR spring rate, but with a preload that works at mid-turn. Then as the weight transfers from acceleration, the RR compresses more than the LR corner and the cross weight will then increase providing more bite off the corners.

**Entry Tuning** – Teams soon learn with bump setups, that they need a much stiffer front shock setting for the rebound of the shocks. They may see this trend of running stiffer rebound shocks as the way to go for all of their shocks. Remember that we set the rate of the shocks in relation to the stiffness of the springs they will be required to control. For the most part, the rear springs are not overly stiff for bump setups.

The left rear corner is usually sprung fairly soft and less than what was used for the older conventional setups. If the spring used in the LR is softer than what was run for conventional setups, then the shock rate in rebound that we use for that corner must necessarily be softer.

Since our spring rate for the LR corner is averaging around 150 ppi with bump setups, our shock rebound only needs to be in that range at three inches per second. Most left rear shocks are valved to a higher rebound setting than that because back in earlier days using conventional setups, the LR spring rate was higher. If we use a shock that is too high in rebound, then on entry to the corner, the LR spring will be slow to extend and weight will be lifted off of that tire and the car will become loose.

This is a very common problem and easily solved by just reducing the rebound in the LR shock. Many crew chiefs I have talked to tell me that the LR shock does not need very much rebound, only enough to control the spring that is installed.

**Putting All Of This to Use** – In order to utilize the bump configurations we have discussed here, we must be able to use a range of different rates of shocks in order to find the right combination for our car at a particular race track for a particular setup. For a team that races at only one track, the process is fairly simple.

For the bump spring setups on asphalt, the whole shock package is much different than when running conventional or soft conventional setups. The bump spring rates will be very high and the shock rebound rates must match those high rates in order to control the bump device.

A bump spring that is in the 1500 pound per inch range will need a shock that is rated from 1200 to 1500 ppi at around 3 inches per second of movement. Usually the ultra-low speed rate of the shock will be comparatively high too and we often see a "nose" rate of between 300 and 600 ppi or more at close to zero speed of movement.

**Summary** – The use of Bump devices has revolutionized setups for circle track cars. For road racing and formula cars, bump springs and bump stops provide a convenient way to keep the chassis from over-traveling on fast and high banked tracks, or from high aero downforce. For those cars, the bumps help to maintain the aero platform and provides consistency.

As for the circle track cars, the setups are becoming much easier to balance, the cars are faster, and the racing is better. What used to be normal in performance differences in a field of race cars is now much less. Race teams have to work much harder to fine tune their setups in order to become better than the competition. In RCT Level Three we will be setting up race cars on bump devices and telling you how to properly do that.

# Exam - In The Context Of This Lesson:

**Bump Setups Provide An Improvement For Which?**

1) Aero downforce
2) Camber change reduction
3) A lower center of gravity
4) All of the above

**Bump Stops Have The Following Characteristic That Bump Springs Don't?**

1) A consistent spring rate
2) A lower spring rate
3) A progressive spring rate
4) More colors available

**Bump Springs Have The Following Characteristic That Bump Stops Don't?**

1) A consistent spring rate
2) wide range of spring rates
3) ore spring travel before coil-bind
4) All of the above

**A Circle Track Sway Bars Can Do What When The Chassis Dives?**

1) A load itself
2) Unload itself
3) Change the cars weight distribution
4) 1 and 3

**The Bump Device Gap To The Shock Body Must Be Timed So That?**

1) The force will be correct
2) It will not coil bind in the turns
3) The weight distribution stays consistent
4) It will load the right front corner first

**Is It Considered Better To Run Bumps On One Corner or Two On Asphalt?**

1) One bump
2) Two, one on each side

## Lesson Seventeen – Anti Dive and Anti Squat

This Lesson on Anti Dive and Anti-Squat is simple and straight forward. We discussed the basics of these Anti's in RCT Level One. What we now have to discuss is how we actually use these Anti's in today's racing environment. Things have changed.

First off, we will talk about Anti-Dive. This was an effect that provided an important function years ago with all race cars, whereas today, the use of this effect is limited by necessity, or sometimes out of lack of necessity. For stock class cars and race cars using more conventional setups, Anti-Dive is still a viable setup option.

*On entry to the corner, a race car brakes and if not restrained, will dive at the front end. This dive can be mitigated and reduced by the effect of Anti-dive. Anti-dive uses the braking force to restrict the movement of the chassis in a downward direction. In this Lesson we will tell you how this is done.*

### Anti-Dive Technology

**What Is Anti-Dive?** – Anti-dive is a mechanical function that utilizes the force of braking to impede the movement of a AA-arm suspension, in most cases the front suspension. For rear AA-arm suspensions, there is an effect of this function that limits droop, or rise, in the rear suspension when braking, but that is very advanced and does not apply to 99% of modern race cars.

So, we have a mechanical function that only works while we are braking. The force created by the brake pads gripping the rotor wants to rotate the brake caliper which is bolted to the spindle is called the torque force. By inclusion, the spindle is also trying to be rotated.

It can't because it is firmly held in position by the two ball joints. But, there still exists a powerful force again that is applied to the spindle while the brakes are applied and the car is being slowed from a very high speed. We can measure, or calculate this force by knowing the torque force created by braking.

There is also the force applied to the chassis and wheel that is the longitudinal force acting through the center of gravity. As the car is braking and the car is slowing, the components still want to move at the same speed, so a high force is created. This force "pushes" on the front chassis mounting points for the control arms and also pushes on the balljoints. The force ultimately acts on, and is centered on, the axle centerline, or spindle pin.

With Anti-dive design, the control arms that attach to the chassis are mounted at an angle from a side view. The upper control arm is mounted so that the rear mount is lower than the front mount. The angle varies, but usually we see between 1 and 6 degrees with 2.5 to 5.0 being the most commonly used angles in stock cars.

*In this photo we see two cars entering the turn. The outside car is much higher under braking than the inside car. It is possible that the inside car has less anti-dive than the outside car. With the bump setups (both of these cars are running bumps), you don't need much if any at all of Anti-dive. We want to get onto the bumps quickly and Anti-dive restricts the movement.*

Most teams split the rates of Anti-dive between the front upper control arm angles in a circle track car that turn only one way. The outside, or right side in the US, upper control arm Anti-dive angle is around 3-5 degrees and the left upper control arm Anti-dive angle is from 1-3 degrees.

The lower arm usually has much less angle and that angle is created by mounting the rear chassis mount higher than the forward chassis mount. These angles are anywhere from 1 to maybe 3 degrees. Now let's analyze what happens with the ball joints in relation to these control arm angles.

If the upper arm is angled the way we explained, then as the spindle/wheel moves up, the position of the upper ball joint moves to the rear. Just imagine rotating the upper arm to vertical, the ball joint would be well behind and to the rear of where it started out if we measured its location from a forward baseline. This motion moves in opposition to the braking torque and in the opposite direction of the longitudinal forces we talked about created by braking.

As for the lower arm, as the lower ball joint moves up with the spindle/wheel, due to the angle, that ball joint will move forward from its original position. It moves much less than the upper ball joint because the arm itself is longer and it has much less sideview angle. This movement is in the opposite direction as the torque movement, and in the same direction as the longitudinal forces.

To sum up the two movements for the upper and lower ball joint, in dive, the spindle rotates counter clockwise when viewed facing the right spindle and clockwise when facing the left spindle as the spindle moves upwards. The motion is in relation to the chassis when the chassis dives.

facing the right front wheel and counter clockwise when facing the left front wheel.

These two opposing motions cause Anti-dive in some degree and the amount depends on two factors:

1) the amount of braking force being applied, and

2) the amount of Anti-dive design angles in the control arms.

If we brake less and have less angle in the control arms, we will have less Anti-dive effect, and visa versa.

*To provide the upper control arm angle needed to produce Anti-dive, offset slugs are used for mounting the arms to the chassis. We would put in a higher hole in the front mount and a lower hole in the rear mount.*

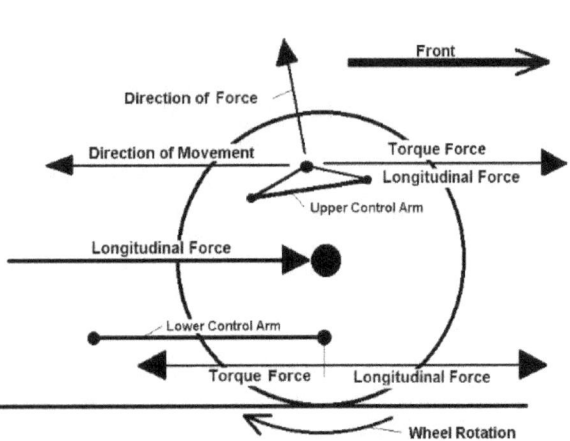

*As this sketch illustrates, as the braking force is applied, the wheel and brake rotor is grabbed by the brake caliper that is mounted to the spindle. This causes the spindle to want to rotate in the direction of the wheel. If the chassis dives, its motion will move the upper ball joint to the rear against the rotation of the wheel. These opposing motions are what cause the Anti-dive effect.*

So, chassis dive will rotate the spindle when we have Anti-dive design in the cars suspension. But as we stated before, the braking force wants to rotate the spindle in the opposite direction, or clockwise when

*Here is what the mounts look like on the car. This car has no Anti-dive as noted by the level control arm mounts. If this car did have Anti-dive, then the front mount would use a slug with the hole higher off the frame rail.*

*Here is a closeup showing the slug and mount. The slug, if offset say ¼ inch and with the hole down from center, could be flipped and that puts the hole up ¼ inch providing a half inch of difference if the rear hole was in the middle of the slot. A half inch is about 5 degrees of control arm angle from a side view.*

**Why Use Anti-dive?** - In earlier periods in racing, and even today with the more stock applications and conventional setups, Anti-dive limits dive in the front end under braking. If we didn't have Anti-dive, then when the car moved down in the front, the cambers of the front tires would change quickly and to a large degree.

Race car tires don't like camber change. We have stated that fact throughout our lessons in the RCT courses. If we can limit the camber change in the tires in a AA-arm suspension, we can create a larger and more consistent contact patch and experience more grip.

So, the race teams that run more conventional setups and the stock classes that by rule must run conventional setups still to this day need Anti-dive. But what about the race cars utilizing bump setups? Good question.

When we run our race car on bumps, the shocks go onto the bumps fairly quickly when we brake into the corner and don't come off the bumps until well past mid-turn. If the shocks are on the bumps, the vertical motion is limited by the very high spring rate of the bump device, whether we are braking or not.

So, if there is very little vertical motion of the chassis while on the bumps, then there will also be very little camber change because camber change only happens with vertical chassis motion and chassis roll, which is also limited with bump setups.

Getting to the point here about bumps setups, if the chassis moves very little, then we don't need to further restrict this motion by having Anti-dive designed into the suspension, do we? No. In fact, we may well want the chassis to dive quickly onto the bumps on entry braking and eliminating Anti-dive would help accomplish that.

As we setup our race cars, we need to think out what we need for effects like Anti-dive. Are we running bump setups? Are we running softer front springs where we might need more help limiting chassis dive? Can and should we control dive through shock settings in addition to the Anti-dive? These are all to be taken into consideration when planning out your setup and the settings that go along with that.

**What Is Pro-dive?** – Some teams have experimented with what is called Pro-dive. This is an effect that contrary to Anti-dive promotes the dive of the suspension while the car is under braking. In this scenario, as the brake force is applied, the angle of the control arms will cause the ball joints to want to move in the same direction as the spindle under braking and helps, or speeds up, the process of dive, again only while under braking.

This author personally does not use nor recommend Pro-dive as a way to promote chassis dive. On entry, the teams that do use Pro-dive suggest that this will move the suspension onto the bumps faster. It may, but in the process, it tends to pull up on the spindle/wheel before actually causing the chassis to dive. This motion lifts load off of the LF tire causing it to lose grip.

With the proper shock rates for compression and the correct soft ride spring, you will have little need for Pro-dive in your front geometry setup. In fact, with those same bump setups, you probably won't need Anti-dive either.

**A Final Note About Anti-dive**  From our lessons on AA-arm geometry, we learned that the elimination of bump steer is partly dependent on the angle of the tie rod. If the spindle rotates while it is moving vertically, and we have designed Anti-dive into our suspension system, then then the angle of the tie rod will necessarily change if the chassis does dive. We will no longer have near zero bump steer after the chassis dives and we would then need to adjust our settings to compensate for the chassis dive influence on the tie rod angle.

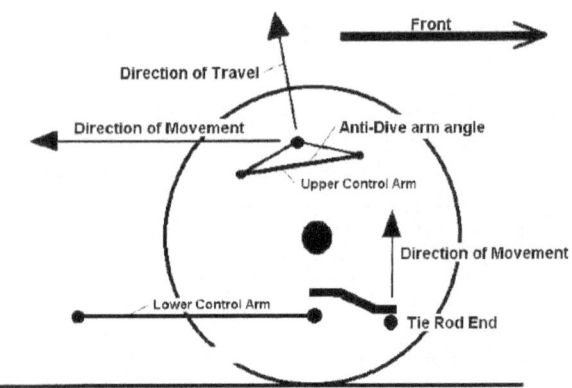

*When we introduce Anti-dive into our suspension system, as the chassis dives, which it will eventually do after we are off the brakes and entering mid-turn, the spindle will rotate from the movement of the ball joints and this rotation also moves the end of the tie rod up. This changes the angle of the tie rod and we might not have near zero bump steer as a result. This has to be re-checked after we change our Anti-dive settings.*

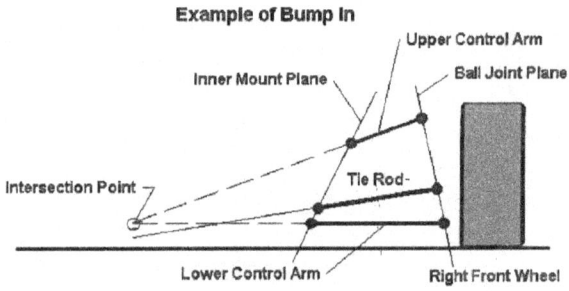

*Here is where we end up after chassis dive when we have Anti-dive present in the system if we don't adjust the bump steer settings. The outer tie rod end at the spindle will move up as the chassis dives and this points the tie rod at an angle that falls below the Instant Center.*

### Anti-Squat Technology

**What Is Anti-squat?** – Anti-squat is a mechanical effect that helps the rear suspension resist squatting when the car is accelerating. Squat comes from the load, or weight, transfer from the front tires onto the rear tires when the car is accelerating.

*On corner exit, we might want to incorporate some Anti-Squat, or Pro-Squat. The car will squat due to the weight transfer that takes place when the car accelerates. This weight can be captured by the Anti-squat links and redistributed between the rear tires for better weight distribution and more grip.*

This effect is very similar to weight transfer during cornering, but it exists at 90 degrees from lateral weight transfer. The same elements are a part of longitudinal weight transfer as are a part of creating lateral weight transfer.

Anti-Squat uses the sideview angles of the upper and lower control arms, or suspension links, to create forces that oppose the movement, or squatting of the rear of the car due to weight transfer during acceleration.

The two forces that assist Anti-Squat are the torque of the motor and the acceleration force pushing the car forward to a greater speed. Both of these forces act on the suspension links. The upper link(s) and the lower links both contribute to Anti-Squat, but we will only concentrate on the upper link for this discussion.

It has been thought that using Anti-squat enhances and adds to the loading on the rear tires when the car is accelerating. That is not exactly true. Like lateral weight transfer, there is a defined amount of weight transfer that happens for the conditions. We cannot change that number unless we change the conditions.

The "conditions" we refer to are:

1) the Center of Gravity height,

2) the weight of the sprung mass of the car,

3) the wheel base length, and

4) the accelerating force, or longitudinal G-force.

As you can see, none of these are easily altered, especially by manipulating the rear geometry in our race cars.

What we can do with Anti-squat is redistribute, some, or all, of the weight that is transferred to the rear and shared among the two rear tires. That is, we can cause more of the weight that has transferred to be placed upon one tire over the other. This can help us move towards more equally loaded tires in the rear and this we have learned provides more overall grip for any pair of tires.

A 100% of Anti-squat condition means that the car will not squat at all from weight transfer due to acceleration. All of the transferred weight is supported by the Anti-squat mechanisms and does not find its way onto the ride springs. We'll tell you how that happens next.

More than 100% Anti-squat will actually lift the rear of the car upon acceleration and lift weight off of the rear springs. This is an extreme example, but some teams have experimented with this effect. It is not advised as a viable alternative to less than 100% Anti-squat.

**How Is Anti-squat Created?** – Anti-squat utilizes the force of the torque of the motor that is accelerating the race car. This torque is acted through the driveshaft, pinion gear and into the ring gear in the rear end. As the car accelerates, the pinion gear tries to climb the ring gear in the rear end. This force is also trying to rotate the rear end in a direction that pushes up on the pinion. From a side view looking at the left rear wheel, the rear end is trying to rotate clockwise.

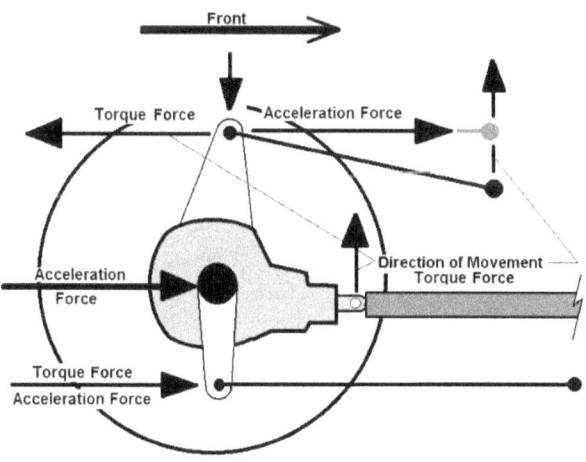

*This shows how under acceleration, the pinion at the end of the driveshaft wants to move up as the pinion gear tries to climb the ring gear inside the rear end. That motion produces a torque force pulling on the third link on top of the rear end. That pulling motion produces a downward force on the rear end at the rear of the link and an upward force at the front of the link that resists squat, hence the term Anti-squat.*

*There is also an Acceleration Thrust force on the upper link mounts that is opposite of the torque force. These two forces must be reduced to a sum of forces to determine the ultimate lifting, or Anti-Squat force.*

The solid axle rear suspension has a system of links that locate the rear end fore and aft from moving when braking and while under acceleration. There is also a mechanism attached to the rear end that is used for preventing the rear end from rotating under those same conditions. In a three or four link system, this part of the mechanism could be either a third link, a lift arm, a lift bar, or torque arm. They basically all do the same thing and that is to prevent rear end from rotating. But these links can also act to distribute the transferred weight differently.

***The Third Link*** - So, we have a third link made of a piece of tubing with heim joints at each end that is mounted to the rear end on the top with brackets. The front of the link is mounted to the chassis and the height of the front mounting bolt is usually lower than the height of the rear mounting bolt. It is this angle of the third link that produces the Anti-squat effect.

*We can adjust the height of the third link at the rear end mount. A lower position, keeping the same link angle, produces more Anti-squat force. It is the distance between the upper and lower control links that determines how much force is produced. We can also raise the lower links to shorten the moment arm.*

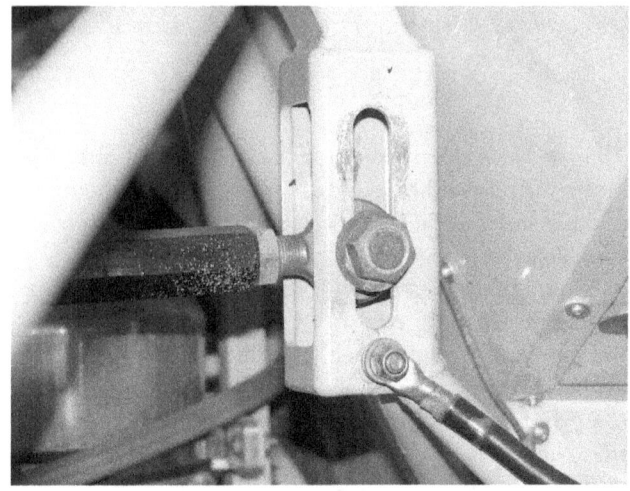

*We can adjust the angle of the third link by lowering the front mount shown here, or raising the rear mount, or both. There is a limit to how much angle we can put into the link for Anti-squat purposes. This is because on entry, the reverse action takes place and there is a tendency for the rear to unload on entry because the Anti-squat mechanism restricts the rear from unloading when the weight is transferred to the front.*

When the race car accelerates, the rear end pulls on the third link with a lot of force. This pulling force is directed at ninety degrees from a line running through the center of the mounting bolt through the center of the axles. Since the front mount is lower and the angle of the third link is not parallel to the ground, it wants to become parallel to the earth and this exerts both a lifting force on the front link and also a downward force on the rear end through those mounts.

For a normal late model racing engine putting out 600 ft/lbs of torque, we can expect a lifting force of about 200 pounds at the chassis for a third link angled at 15 degrees. If 300 pounds transferred to the rear under acceleration, then the chassis would only squat 1/3 of the amount it would have if there were no Anti-squat. A full 2/3 of the squat loading would be absorbed by the third link angle.

Now comes the best part. That 2/3 of the transferred weight, or about 200 pounds, now is placed onto the rear end instead of onto the ride springs. The center of this force is positioned at the center of where the rear of the link is located side to side on the rear end. If that link were equidistant from each of the rear tires, then 100 pounds of load, or half of the total, would be placed on each of the tires. If this 200 pounds of load were transferred to the springs instead, it might be distributed differently depending on the spring rates of the ride springs.

Now, if we move that third link mount towards one of the rear tires, then that tire would have more of the transferred percentage of weight added to it. Let's say we move the third link 6 inches to the left on a rear end with a 66 inch rear track width. Then the percentage of LR added weight of the 200 pounds of available transferred weight would be 118 pounds instead of the 100 pounds we added when the link was centered between the rear tires. Not only is the 18 pounds added to the LR tire, 18 pounds is taken off of the RR tire. This represents a difference in loading of 36 pound for the two rear tires from what it was before. That 36 pounds equals 1.4 percent of the total vehicle weight.

*The Anti-squat forces created by the third link can be used to place some of the weight that has been transferred from front to rear under acceleration. If we move the upper link closer to the left tire, then that tire will receive more of the transferred weight than the right tire.*

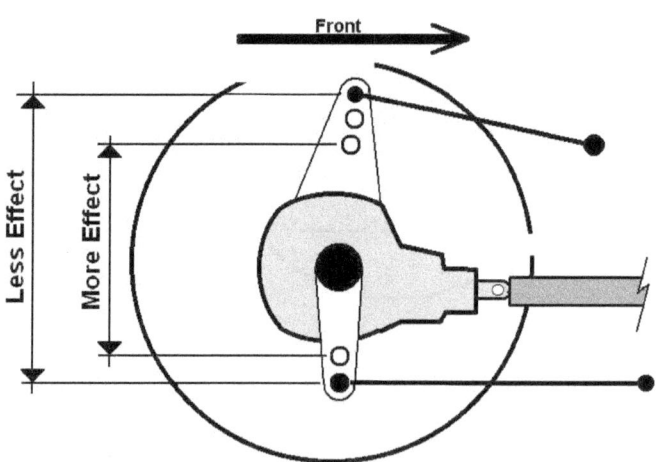

*If we lower the upper link in a three link suspension and/or raise the lower link, the distance between the upper and lower mounts becomes shorter. This shorter distance provides more Anti-squat force for more effect. The opposite produces less effect.*

***The Lift Arm, Torque Arm and Lift Bar*** – These devices are designed to prevent the rotation of the rear end while the race car is under acceleration as a first priority. Secondly, they act much like the third link and transfer some of the acceleration weight transfer onto the rear end.

These devices have torque softening springs and shocks to help absorb the application of torque when the race car is throttled up on corner exit. But those don't affect the Anti-squat properties. What does affect those properties and how much Anti-squat is present is the length of the arm from the centerline of the axle to the front chassis mount.

*This example of a Lift Arm is used on a dirt late model. Some asphalt late models are now experimenting with lift arms too. Note that the rear of the arm is mounted to the rear end and has no lateral adjustment to move the force towards on tire or the other. The position the coil-over is mounted can be adjusted to make the arm longer or shorter.*

If we shorten this distance, then the greater the force of the torque that is applied as Anti-squat. If we lengthen the arm, then less force is applied to the chassis and rear end. Lift arms and such are not easily adjustable for lateral location on the rear end like a third link is. Therefore, where the load absorbed by the lift arm, etc. is applied is mostly fixed.

**What Does Anti-squat Do?** - Anti-squat transfers some or all of the weight that has transferred from the front tires to the rear tires off of the springs and onto the links attached to the rear end. The law of physics state that for every reaction there is an equal and opposite reaction. This means the Anti-squat effect cannot add load onto the rear tires. The total loading that would normally be there if we didn't have Anti-squat remains the same.

This is understandable because where would the added load come from? In the study of dynamics, we cannot "receive" additional load without "taking" from somewhere else the same amount of load. So, we receive the loading on the rear end from the action of Anti-squat and that same load is taken from the rear ride springs where it would have ended up if we didn't have the Anti-squat effect.

Let's look at it another way. If we had zero Anti-squat effect, the car would transfer a certain load or weight onto the rear springs as we accelerated and they would compress to carry that added load and the car would squat. If we apply 100% Anti-squat meaning the effect would prevent the car from squatting at all upon acceleration, then the springs would not have compressed and therefore do not carry any more load. The load the springs would have carried is now applied to the rear end through the Anti-squat device because it has to go somewhere.

**Adjusting The Amount/Location Of Anti-squat** - There are several ways to adjust the magnitude of the Anti-squat vertical force that is applied to the rear end through the device. One is by changing the angle of the third link and the other is by changing the distance that the upper link is mounted from the lower links. The greater the distance between upper and lower rear mounts, the less mechanical force the system has and therefore the less Anti-squat effect.

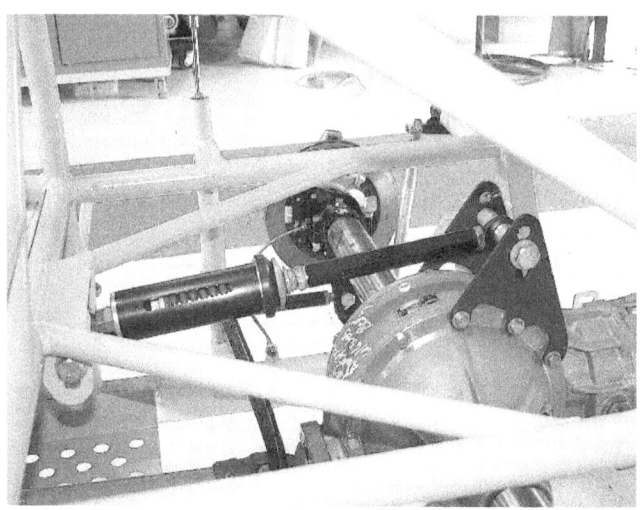

*This third link has a spring that compresses when the car accelerates. The motion that help produce Anti-squat is the same force that pulls on the spring. This cushions the movement, but does not restrict the Anti-squat effect.*

Adding angle to the third link will increase the Anti-squat effect but this has its limits because the effect is reversed when we decelerate and brake into the corners.

Instead of transferring load onto the rear tires, the reverse means that we reduce load on the rear tires when entering the turns and that can make the car very loose.

Usually an 8 to 10 degree angle is sufficient to cause the desired effect while not being excessive. If more Anti-squat effect is needed, we can lower the third link and/or raise the lower trailing arms while maintaining the angles of all of the links.

*This close-up shows how with the third link, there is plenty of room to move the rear heim of the link to the side. Moving it more to the left, and closer to the left rear tire, will cause more loading of the left rear tire from the weight that transfers while under acceleration.*

As we have stated, with Lift Arms, etc., we can adjust the length of the arm to change the amount of Anti-squat we have. The shorter the arm, the more force we will have for Anti-squat.

**Summation** - Anti-dive is useful for setups that do not utilize bumps. These include stock classes of race cars, road racing cars, and those with more conventional setups like late model circle track cars.

In other Lessons we discuss how to geometrically steer the rear end by utilizing the squatting action of the rear of the car. For this to happen, we need for the car to squat to a certain degree. If we ran 100% Anti-squat, then there would be no squat, and therefore no motion that would have the desired effect for rear steer.

Anti-squat, as we have said, does not produce more loading on the rear tires under acceleration. What it does is restrict the rear from squatting and can be used to re-distribute the weight that has transferred due to acceleration. It is not the magic bullet so to speak, but it can be used to our advantage if setup correctly. We have given you some ideas on how to do that.

# Exam - In The Context Of This Lesson:

### Anti-dive Works When?

1) When the race car is under acceleration

2) As the car goes through mid-turn

3) While the car is braking

4) As a product of roll center location

### With Anti-dive, Under Braking The Ball Joints Move?

1) Opposite of the braking force

2) In the same direction as the braking force

3) Towards the inside of the turns

4) Towards the outside of the turns

### Advantages Of Anti-dive Include Which Of The Following?

1) Less camber change

2) Reduced over-travel of the front end on braking

3) Better grip on corner entry

4) All of the above

### Anti-dive Should Not Be Used For?

1) Bump setups

2) Stock class setups

3) Conventional setups

4) Road racing setups

### Pro-dive Should Only Be Used For?

1) The right front corner

2) The left front corner

3) The right rear corner

4) The left rear corner

### Anti-squat Cannot Not Do What?

1) Reduce chassis movement when accelerating

2) Redistribute the loading on the rear tires when accelerating

3) Reduce the loading on the ride springs under acceleration

4) Add loading to the rear tires under acceleration

### To Increase The Anti-squat Effect, We Do What?

1) Lengthen the lift arm

2) Raise the third link

3) Increase the third link angle

4) Lengthen the third link

### 100% Anti-squat Does What?

1) Reduce chassis movement to one-half

2) Puts all of the rear loading onto the rear end

3) Provides 100% of rear grip potential

4) Eliminates all rear movement under acceleration

### Moving The Third Link Left Does What?

1) Increases the effect of anti-squat

2) Puts more loading onto the RR tire under acceleration

3) Puts more loading onto the LR tire under acceleration

4) Loosens the car coming off the corner

### To Enable Rear Steer, We Need?

1) The most anti-squat we can get

2) Less than 100% of anti-squat

3) A higher third link angle

4) A lower third link location

## Lesson Eighteen – Ride Heights

Every race car from dirt cars to Formula One cars have established ride heights. These heights provide the basis for every vertical and angular measurement we make throughout the life of the race car. Changes in ride height that go un-noticed will change control arm and suspension link angles and cause chaos.

We can either create and note forever our ride heights, or we can go by the original manufacturers suggestions. In any event, we first off need to establish what ride heights we will use and then return the car to these established ride heights whenever we change a setting or part.

*We can measure our ride heights to the floor from raised stands. These stands are the same height at each corner of the car and allow us to move around under the car to take measurements of control arm angles, and other geometric functions.*

**Ride Height Definition** – I guess it would be a good idea to define just what constitutes a ride height. For most cars, we would choose a point on the chassis near each tire that we can easily get to in order to measure from the ground to the frame. We can re-create these heights later on using other measurements in places that may be easier to get to.

These heights may be dictated by the rules you are governed by, or what makes sense in your class of racing. There are many ways to measure ride height. For formula cars that have no easy place to measure from, the height can be dictated by the length of the spring/shock mounts. These are usually on top of the chassis and easier to get to.

For those cars, once the correct suspension angles are achieved, and the correct heights of other components that matter are achieved, the team will simply measure the shock/spring length and then go back to that measurement in the future. That measurement of the shock/spring length relates directly to the chassis height.

For most circle track cars and road racing cars, the heights are taken from the chassis to the ground. Like the formula cars, we can translate these measurements to other suspension parts to make life easier later on, like when at the track.

**Ride Height Considerations** – The most important thing to remember about measuring ride heights is to make sure the surface on which you are measuring is flat. I didn't say level because we can get away with a small amount of slope. The spots the tires rest on must be on the same plane or our ride heights will change once we run the car on the relatively flat plane of the race track. If your floor is not level, then the slope amount must be under an inch in the length of the wheel base and a half inch in the span of the track width of the tire centers.

*To get a surface for our four tire contact patches that is on the same plane, we use a level across the spots at each end of the car. We use spacers under the low spots to bring them up so that each pair of contact spots are on the same plane, or level to each other. The floor can have a rake angle as long as it is not excessive. Up to a one inch difference from level in the length of the wheel base, or up to a half inch difference from level in the track width is acceptable.*

To make sure your tire spots are on the same plane, you can just place a level across the centers of each tire location at each end of the car. So, the front tires should be at the same angle as the rear tires. I didn't say level here either, but the difference must be under a half inch of height difference from level over the track width, or distance between the tire contact patch centers.

The best thing to use is a six foot level obtainable from any hardware store for under $50. Or, you can use a length of tubing, square metal, stiff piece of wood, or whatever works to use to span the distance and measure that with a smart, or bubble, level. Shim the low end until the level is, well, level. Then use those same shims at the other end. To create a "same plane" condition for the four tires, find the lowest tire location and place plates/spacers under that tire, or scale if you are using scales, until the tool is level. Then your four tires will be on the same plane.

Example: Let's say the front level is established and the shims are under a half inch. Then we move to the rear and we find that we must raise one end of the tool to get to level again. Simply place plates of thin metal, or even layers of paper or plywood under that end of the measuring medium until the level shows it is level. Use those same spacers under that tire, or if scaling the car, on top of the scale for that corner every time you setup the car or change the weight distribution.

**Spring Changes** – We will cover this section in more than one Lesson, but it helps to drive the point home sometimes. When making a spring change to a new rate of spring, or a new length, the ride height will change. If you are installing a softer rate of spring in one corner of the car, that corner will be lower unless you adjust the spring loading back to where it was by use of an adjuster.

The adjuster can be a ring on a threaded shock body, a large threaded bolt placed on top of the spring mount, or other means, but in any event, something must be adjusted in order to get the ride height back to what it was before the spring change.

*For big springs that are used in stock class applications, we can use an adjustable spacer to achieve the ride heights we desire. If the class you run in does not allow adjusters, these can be welded solid after you find your desired ride height. Then you will be legal. Just remember that if your change spring rates, your ride heights will no longer be what you had originally.*

*When we change spring rates on one corner of the car, the ride height will change. So, we need to adjust the spring height so that our ride height will return to the original ride height. Installing a spring with less rate will cause the ride height to be less. We then adjust the spring height up to compensate. Here the crew member is using the ring adjuster on a coil-over shock/spring combo.*

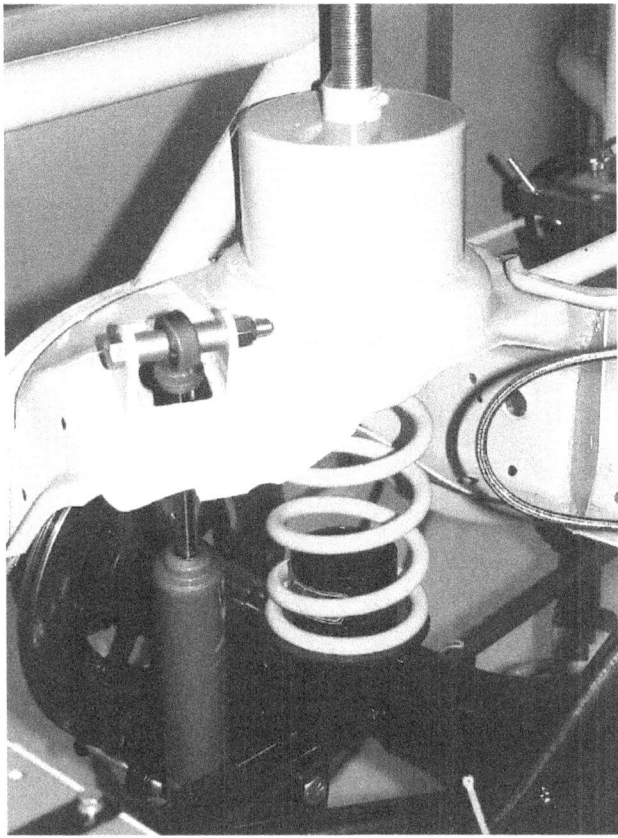

*For this application, the car builder installed screw jacks over the springs to easily adjust the ride heights. These are typically used on the front of stock clip (stock front frame rails) that use stock sized big springs.*

If you want to make the job of changing a spring easier, just change one spring at a time. Then you will only need to make one ride height adjustment at a time. Once you are done with that corner of the car, go to the other corner or corners you want to make a spring change for. But do them one at a time.

**What Height Do I Need** - For circle track cars, tradition has it that there is usually a rake, or difference in ride heights among the four corners of the car. The Left Front is usually the lowest corner with the Right Front next followed by the Left Rear (sometimes the same as the RF) and then the Right Rear being the highest ride height.

With the modern setups, these "norms" may not be so normal anymore. Teams and manufacturers car re-thinking the ride height layout to accommodate the newer low attitude setups and making the ride heights so that the car has less rake and is more level, like it will be on the race track.

For road racing cars, there can be a rake, but the corners at each end of the car are usually the same height. So, the front would be the lowest ride heights on those cars. Much of the decision for ride height on road racing cars is connected to rules governing wing heights, or front valance heights, etc.

The road racing cars might have a flat bottom where they need to maintain what is called a platform. The platform is the angle and height of the flatter bottom of these cars to the racing surface. For aerodynamic purposes, the dimensional heights of the platform are critical. Small changes to the ride heights can cause large gains or losses in the aero downforce that are created by the platform.

**Tools To Measure Ride Height** – There are various tools racers use to measure ride height. These range from a simple tape measure to sliding blocks that fit between the frame and the ground. Once you establish your ride heights at the shop, you can recreate them at the track by recording the shock lengths and making sure they are still the same after the trip to the track if, and only if, the spot you park your car creates the plane we used at the shop.

NOTE: The spot you park your car at the track is seldom flat or to where the four tires are on the same plane. Therefore, you will need to either shim the spots where the tires rest, or record new ride heights that you can use to maintain the geometry of your car. These won't match the ride heights you established at the race shop, but will maintain the heights once the car hits the track.

Even if the tires rest on uneven spots that are not on the same plane, when you make spring changes, if you keep the same "track" ride heights, your overall geometry and static tire loading will not change when the car is on the "same plane" track surface.

**Ways To Record Ride Height** – There are several, or multiple, ways you can record ride height. You can just measure from the corners of the frame to the ground, and this is by far the most common. You can also measure the shock length at each corner, or from the bottom of the wheel rim up to a point on the body for full bodied race cars.

*Once you have established your chassis ride heights by measuring from the floor to the frame, you can record other dimensions to replicate those ride heights. These alternate methods of measuring are many times much easier to get to and perform. One way is to measure from the bottom of the wheel rim to the fender well.*

*Another way to replicate ride height measurements is to measure from the ball joint up to the frame tubing or measure the shock length. Either of these will place the chassis at the same attitude that the ride heights created initially.*

For open wheel cars, you can measure from the top of the lower ball joint up to a point on the chassis. Whichever way you choose, be sure to record those measurements so you can access them later on. Some teams just write them on a piece of tape attached to the body or frame so anyone can read them easily.

**Other Uses For Ride Heights** – We have already covered some uses for ride height. Just making sure our ride height don't change when making spring changes is the most common use. But we can use them for other things.

If we make careful and precise measurements of the corners of the car, then after a hard hit, we can remeasure those corners to see if the frame is bent. A bent frame will alter our component angles and change our setup.

If we have a need to make many measurements, such as for geometry purposes, and we need to get under the car, we can add a constant offset measurement to the ride heights at all four corners to raise the car, but maintain the rake as if the car were on the ground.

After we make the measurements, we can then just subtract the vertical offset to record the component heights as if the car were on the ground. If say, our ride heights were LF = 4.0", RF = 4.25", RR = 4.5" and LR = 4.25", and we add 10.0" to raise the car, what would our new heights be? They would be LF = 14.0", RF = 14.25", RR = 14.5" and LR = 14.25".

**Alternate Ride Heights** – Sometimes we want to take measurements on the car that will relate to how it is situated on the race track at, say, mid-turn. This could be called the dynamic ride heights because they are caused by the dynamic forces as the car rides through the turns.

*We not only need to know our static ride heights, we also want to know our dynamic ride heights. These are the heights the chassis is at when the car is at mid-turn and all of the forces are on the car. We can record the dynamic heights using data acquisition and position sensors mounted to the shock, or by other means, but the goal is to know how far each corner of the car travels at mid-turn so we can duplicate that travel in the shop.*

If we know the shock travels from reading the travel indicators on the shock, then we will know how much shorter our shock mounts are at dynamic ride heights. If we use adjustable solid links to replace the shocks, we can adjust them to the new shock length, place them in place where the shocks are normally mounted, and then place the car on the ground, or better yet, onto raised stands.

*Simple shock travel indicators are mostly used in many different forms of auto racing to record the maximum travel of the springs, and with those, the suspension. We need to be careful when relying on those measurements. Maximum travel might also include the dive the front suspension experiences during heavy braking, or the squat the rear experiences during acceleration off the corners. The team can make a run reaching full turn speed without heavy braking or hard acceleration to get these travels to represent just the chassis movement related to the mid-turn forces.*

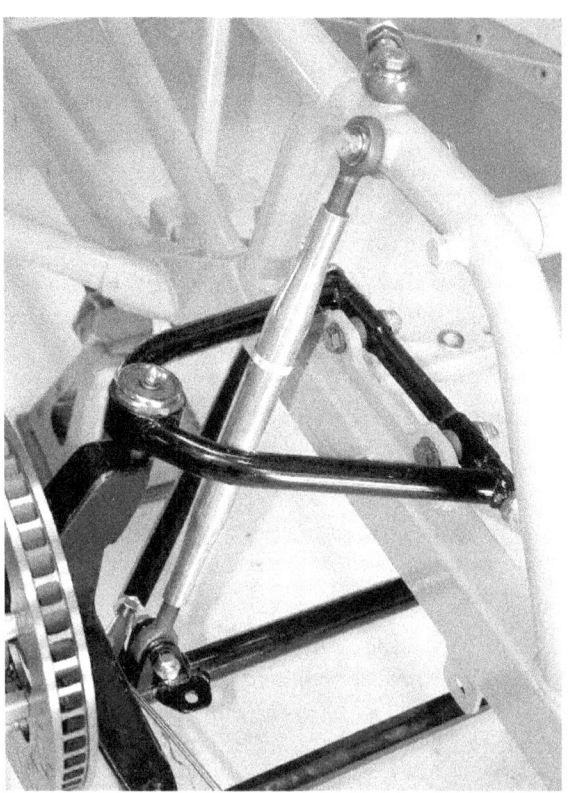

*Once we know the shock lengths from dynamic shock travel, we can install links that are the same measurement so that the chassis will be at dynamic ride height. Once there, we can measure front AA-arm angles, link angles in the rear, wheel cambers, clearances, and other important features. To be able to measure tire loading while at dynamic ride height, we need to use a pull-down rig that actually loads the chassis and springs. Then we can record the dynamic loads seen on the scales.*

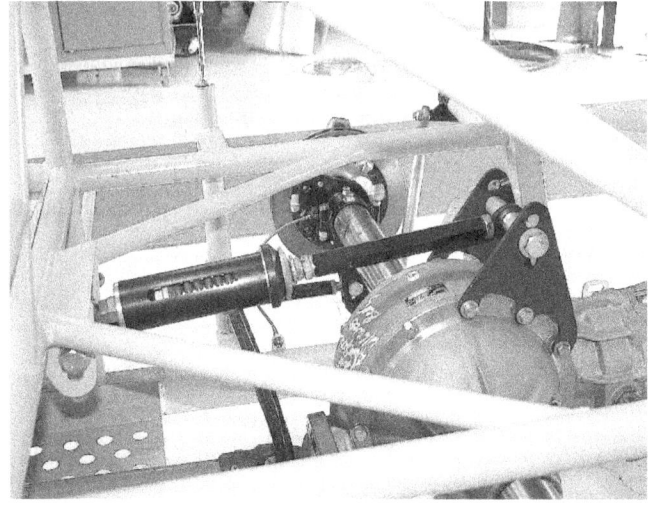

*Pulling down the chassis to dynamic ride heights will tell us how our control arm angles are positioned, what our driveline angles are, and what our front wheel directions are. All of these might change under loading as opposed to just placing the components at that attitude with no loading.*

We might have many needs to put the car at dynamic ride height. Some of those include: measuring control arm angles, measuring wheel direction, or steer, measuring frame clearance to the racing surface, measuring driveline alignment, etc.

We now have available what are called Pulldown Rigs. These will pull the chassis down to the ride heights that the car has traveled to at mid-turn. Then if the car is on scales, the team can read the loading on the tires, as well as geometric measurements that reflect where the components are under the stress of dynamic loading. For instance, the load on the chassis and control arms from loading might alter the toe readings.

**Summary** – It is important to establish and record ride heights for your race car. You can take the measurements in a lot of different ways, so choose the one that best suits the design for your race car.

Make sure the surface where you will be setting ride heights in your race shop where the four tire contact patches will sit will all be in the same plane. And the overall tilt of that plane should be no more than one inch in the wheel base length and a half inch in the width of the track, or tire width.

# Exam - In The Context Of This Lesson Three:

### What Is A Ride Height?

1) The maximum height the car will achieve on the track

2) The distance from the frame to the ground

3) The distance from the body to the ground

4) 2 and 3

### We Maintain A Defined Ride Height Because?

1) The suspension angles will always be consistent

2) The static loading on the tires will stay consistent

3) Aero downforce will be predictable

4) All of the above

### The Surface We Measure Ride Heights To Must Be?

1) Flat

2) Level

3) On the same plane

4) The same as at the track

### We Can Maintain Our Ride Heights By Measuring What?

1) The distance from the floor to the corners of the chassis

2) The shock lengths

3) The ball joint stud to the frame

4) All of the above

### When We Get To The Track, We?

1) Re-set our ride heights to what they were at the shop

2) Level the spots under the tires

3) Record new ride heights without changing tire loading

4) 2 or 3

### When We Change A Spring, We?

1) Make sure we don't change the spring adjuster

2) Return to the original ride heights

3) Return to the original corner weights

4) 2 and 3

## Lesson Nineteen – Tire Selection and How To Read

Everything we do with race car setup all comes down to one thing, making the tires do as much work as possible. In the very first introduction to Race Car Technology we said that our goal is to: "Maximize the amount of traction that is available from the four tires on a race car, any race car, and that will make you as fast as you can be, all other things being equal. Everything we present will ultimately lead to optimization of the race cars tires grip and we use of that grip to go faster."

*The art of the selection of race tires and the preparation and maintenance are critical to your cars performance. Everything we do for chassis setup is directly related to helping the tires work harder. This Lesson will provide you with some ideas for choosing and preparing your race tires.*

In order to maximize the grip in our tires, we need to know how to select those tires, how to prepare the tires and how to read them once they are run on the race track. Remember, the goal is to produce the ideal loading among the four tires and create the largest contact patch possible. Then our race car will be as fast as possible through the turns.

Every race team should select a tire specialist or a small team of specialists to choose, prepare and maintain the teams race tires. Success depends on it. That teams needs to keep good notes and go over those notes often. The following will help those guys with their duties.

**Tools Of The Tire Trade** – The tools that are considered absolutely necessary are:

1) a quality, large dial pressure gauge,

2) a small, flat tape measure,

3) tire chalk or suitable marker,

4) extra stems, bleeders, valve cores, etc.,

5) a note book you can use to record your tire sizes and pressures,

6) and a nitrogen tank to use to pressurize your race tires.

Other not so necessary tools, but ones that can help you, include a durometer to test the hardness of your tires, a tread depth gauge, purging equipment to remove most of the atmospheric air, that contains moisture, from your tires and wrapping material to keep your tires from drying out when not in use.

You can also invest in tire covers for use at the track when the hot sun is on one side of the car heating those tires when the other side is in the shade. Some teams also carry tire brushes to wipe off the sand or pebbles before measuring the sizes so that the numbers come out correctly.

*Use a large diameter tire pressure gauge so that you can read, and estimate values between the whole pound marks. We need to work with half pounds of tire pressures in many cases.*

*Tire temperatures tell us a lot and help us to apply the correct tire pressures as well as cambers. When you take your temperatures, poke the probe into the rubber at a 45 degree angle and push it close to the core. The surface temperature cools very quickly and won't help us understand the true heat the tires generate.*

*Measure your tire diameter around the middle just off centerline to avoid the seam. Measure the tire un-inflated first to get relative sizes, then with race pressures plus 10 pounds. The chalk marks on new tires are the sizes of the tires inflated to the same pressures plus some amount beyond race pressures.*

**How To Select New Tires** – The overall theory on tire selection deals with a few critical musts. They are:

1) The tires must be as fresh as possible,

2) They must be equal in age,

3) And for circle track race cars, they must be sized correctly so that you can mount the correct stagger on the car.

*Most race tracks will provide a stack of race tires for each division. It is up to the team to dig into that stack and find the correct tires that will work well for your race car. It is the age, sizes and grouping of the tires that matters the most.*

*For Circle Track Asphalt Applications* - When you go to the tire shack at your track, you are looking for sizes and dates. Let's deal with the sizes first. The manufacturer measures all new tires that have been inflated a few pounds above the highest race pressures. They write these sizes on the tire in chalk. These are referred to as chalk marks.

Since all tires, left and right, are inflated to the same pressure, naturally, when you mount your tires and inflate them to different left and right side pressures, the left sides will be smaller due to the lower pressures. So, choose a size difference that is much smaller than your required stagger.

Measure all of the tires in their deflated state. Measure the tires around the middle just off center to avoid the seam. Keep the same sizes for the right side tires and keep the same size for the left side tires. The stagger you are looking for with the deflated tires is about half what you need on the car when inflated to race pressures.

So, if you choose an 85 inch tire circumference for the right side tires, then if you need 2 ½ inches of stagger, you would choose 83 ¾ inch tires for the left sides, because the inflated stagger will be twice as much. This number may differ for different tires at different race tracks, and only repeated selection will determine the amount of difference in stagger between the deflated and inflated tires that you will end up using.

The date codes are molded into the tires and printed on the sticker. It is critical to match the date codes for each side of the car. That means, the right side and left side tires don't necessarily have to have the same date code, and chances are, they won't. But the RF and RR tires, as well as the LF and LR tires must have the same date codes or the car will not be balanced and the handling will suffer.

This will take some work and sorting through the various tires that are mixed and un-matched lying around your tire shack, but the top teams make the effort to get this right. It should be done as soon as possible and it should take as much time as is necessary to get the right tires, period. Get this wrong and you'll be out to lunch for as long as you have those miss-matched tires on the car.

*A hard tire does not adhere well to a rough surface because the contact patch does not conform to the entire racing surface area. The total square inches of rubber that meets the track is reduced when the tire is harder. Older tires, even if they have very few laps on them, will be harder. The chemicals that make the compound pliable evaporate over time. Sun and heat accelerate the process. If legal, a team can apply commercial chemicals made for racing applications to the tire to make them more pliable.*

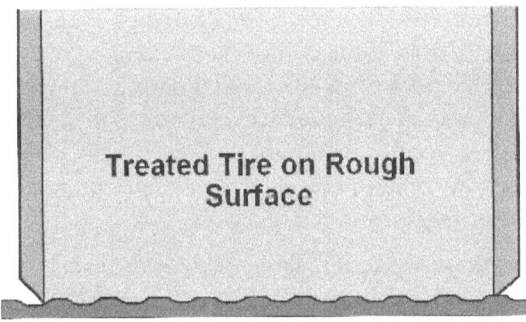

*Tires have more grip by being newer, or by being treated. By treating the tires, the surface is changed chemically so that it is more compliant and "fills" the areas around the high spots in the track surface. A treated tire would provide a performance benefit nearly as much if the track was very smooth.*

*Storing your race tires in a plastic wrap helps keep them fresh and delays the evaporation of important chemicals that make the tire soft and compliant. This is true for normal or treated tires.*

*Road Racing Tire Selection* – Most road racing series and sanctions choose the tires for the teams. If you have a choice, then go by the date codes and choose all the same dates for your four tires. If you are choosing multiple sets of tires, make sure to practice on the older dated tires and save the newer dated tires for the race. This last statement is true for all racing types.

If there is a difference in tire diameter sizes and all of the tires you use are the same width, choose the same size for each end of the car if you cannot get the same size for all four tires. Put the smaller diameter size on the front when mounting the tires.

If your type of racing uses different width tires for the front and rear, choose you tires so that the date codes are the same (might be hard to match) or as close to the same date as possible. For instance, if you choose different dated tires for the front and rear in two sets, and the two sets are reversed (older fronts on one set, newer fronts on the other), then the handling between these two sets of tires will be vastly different.

**How To Read The Sticker** – A new race tires has a sticker on it that tells certain important information. Learn how to read your tires sticker. If you need to, have a talk with the tires representative at the track, or call the company. They should be glad to help you with this information. The information includes the date the tire was manufactured, the group or batch it was made with and other information about the tire compound codes, etc.

*The stickers on new tires can tell us a lot about the age of the tire. They help us select tires that are the same relative age, which is important maybe more so than the age. If we cannot find left side tires that have the same dates as the right sides, at least match dates for front to rear tires on the same side. That way we will maintain the balance of the car.*

For a Hoosier race tire, the sticker has information you need. The top number is the model number and is usually five digits. It might say 10615. The next number is the tire size. This will look like 27.0/10.0-15. Under that number is the compound number and it might look like F40.

Then there is a bar code and just under that is a date code and sequence number that no one can decipher. So, you need to try to find tires with the same sticker date number, and that might be very hard to do because a production run of a certain tire might last several days at the factory.

What is the same, for the same batch is the date code embedded in the tire sidewall. This code will be the same for all of the tires coming out of that mold in that production run regardless of the date and sequence. When that run is finished, they will change that code for the next run which usually involves producing tires with a different tire compound and/or core design.

**Mounting Your Tires** – Make sure you mount the tires on the correct rims. Some teams will run different wheel offsets for each side of the car. Make sure the imbedded date code for each tire is on the left side, or infield side of the rim for both sides for circle track cars. So, on the left sides, you'll be able to see the code when mounted on the car and on the right sides, you won't.

Don't allow the use of any liquid like bead soap, or use very little if needed. Liquid that gets into the inside of the tire will expand when heated and influence the tire pressures when they heat up on the track.

Purge the tires with nitrogen by inflating and deflating two or more times. Inflate your tires to about ten pounds above the race pressures and measure them.

Then deflate them to race pressures and let them sit. Once they have stabilized, then re-measure them at the race pressures.

This is the point in time where you create your sets of tires from a group of eight or more. The sets will consist of tires at each end of the car that will have the correct stagger for your track. If a tire needs to "grow", you can inflate to 10 pounds above race pressure and then let it sit for a while and that may increase the size slightly.

*This tire has been marked for the corner of the car, the set number, the track is has been run on, and the date. In chalk or other less permanent style, the size has been marked as well. The size can change with pressure changes and/or ambient temperature changes. That is why we mark them with less permanent markers so we can change the number.*

Mark the tires as to the corner they belong on, the set number, the track it was run at and the date. In your notebook record each tires size and pressure. As you run the tire, record the number of laps run also. Race tires, like any other tire, will degrade in performance the more they are run and the older they are.

**Safety Note:** Never inflate the tires more than 10 pounds above race pressures. Do not ever use air hose ends that can be clamped onto the valve stem. This is a recipe for disaster. If you become distracted, the tire can over-inflate by accident and explode with the power of a hand grenade. Race team members have died using these types of clamp hose ends.

*Never inflate your tires with an air hose that clamps onto the valve stem. This is a very dangerous practice and if left unattended, the tire can overinflate and explode with the force of a hand grenade. Racers have been injured and killed doing this.*

**Choosing Stagger** – Even when we know our desired stagger amounts, we can mount the tires such that changes can be made to improve the car due to unexpected stagger changes. Once the tires get hot, the gas expands and the bias tires will stretch. If we put the correct stagger in the rear, we can set the front stagger so that we can swap tires on each side of the car to effect the rear stagger if one of the rear tires changes its size unpredictably after being run.

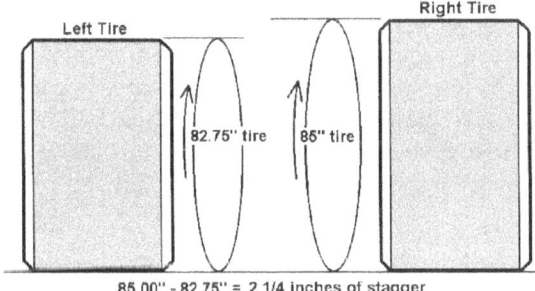

*Stagger is the difference in circumference of two tires on the same end of the car used for circle track race cars. The example is for a car that turns left. Measure the stagger around each tire near the middle of the tread. Subtract the two measurements to find Stagger or "Roll Out" as it is sometimes called. Always inflate to race pressures before measuring.*

*To find stagger the easy way, follow this simple method and never subtract tire sizes again. Measure the larger tire first. Then remember the number.*

*Wrap the tape measure around the smaller tire and note where the larger tire stagger number falls near the end of the tape. Read the measurement from the end and you have the stagger.*

Suppose we need a rear stagger amount of 2 ½ inches. We mount a RR tire that is 86.0 inches in circumference and a LR tire that is 83.50 inches around. A common occurrence is when one of the rear tires will grow excessively causing an increase or decrease in stagger. If we mount the same stagger on the front, but at a larger size, we can swap either left side or right side tires to correct the problem.

The tire stagger must match the race track and groove radius we will be racing on. We do not want to correct handling problems with excess or deficient stagger. For every track there is an optimum stagger for the rear tires and if we use a locked spool rear differential the need for correct stagger is even greater.

The spool will need a stagger that is an average of the radii that the tire will experience in the turns. Very few driving lines result in a single constant radius. The driving line through a turn is more like a parabola, or constantly changing radius with the smallest radius in the middle portion of the turns.

A Detroit Locker rear differential unlocks going into the turns and locks back up upon acceleration. For this type of rear end, we need to match the stagger to the radius of the last third of the turn. This may be less stagger than what we might have needed at the very tightest portion of the mid-turn.

**Pressures For New Tires** – The following is for circle track cars mostly, but much of this information is relevant for road racing tires too. Starting pressures for new, cold tires will be lower than race pressures. The left sides should be about 4-5 pounds lower and the right sides 7-8 pounds lower. Run the tires for one practice session and then adjust the pressures and check the stagger. Make changes based on the tire temperatures and hot pressures as necessary.

If you are allowed to run bleeders, don't run them in practice so you can see what the pressures are doing. This can identify a handling problem that would otherwise go un-noticed, such as a growing RR tire due to being loose, etc.

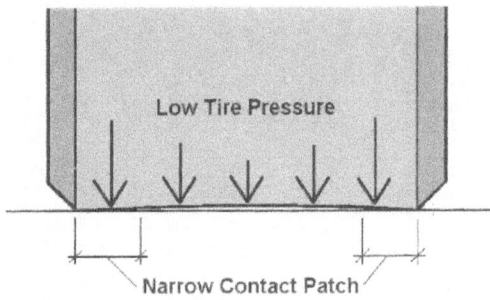

*It's a the fact that the whole area of the contact patch is not exerting the maximum pressure on the racing surface, and that is a the problem. This tire will have less grip than one that is properly inflated. Here we see low pressure where the center of the contact patch is not providing maximum loading due to the pressure being insufficient to press in onto the racing surface.*

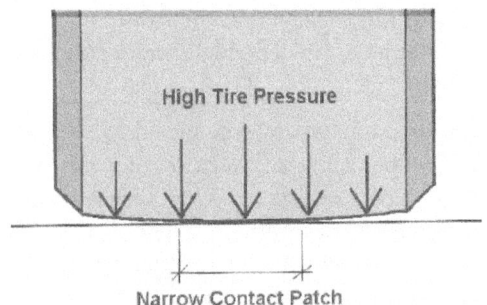

*Using excessively high tire pressures will also cause the tire to have a smaller contact patch area and therefore reduced traction. It is not that all of the width of the tire is not on the ground. Here we see excessive pressure puts most of the load at the center of the contact patch instead of distributing it across the entire tread width.*

**Tire Temperatures** – It is very important to take tire temperatures when evaluating your cars setup. The temperatures can tell us how well each tire is working in conjunction with the other three tires. We set camber and tire pressures based on how our tire temperatures look.

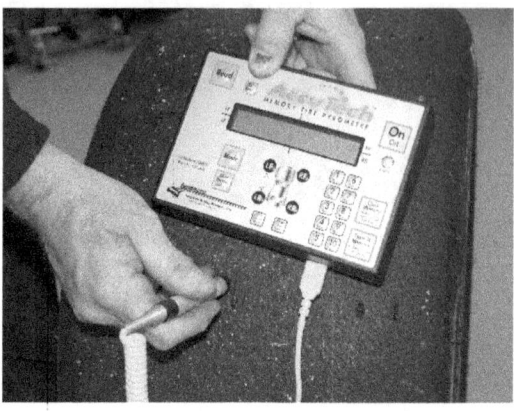

*This shows a tire temperature gauge while the crewmember is taking a reading. Take the outer edge temperatures in from the outside edge of the tire approximately 1 to 1 ½ inches first and then the middle and then the inside edge. Allow enough time for the reading to settle before recording each number.*

The proper way to take tire temperatures is to read the temps in three places across the face of the tire with a tire temperature probe. This instrument has a needle like end that is inserted into the rubber to get the inside, or core temperature. It should be inserted at about a 45 degree angle into the rubber. The temperatures for the three positions should be taken near the inner and outer edge of the tire about one inch in, and at the middle of the tire.

Give the instrument time to react to the temperature of the tire. Once the reading stabilizes, record the amount and move on to the next reading. Don't be in too much of a hurry to do this because a lot of decisions will be made based on the temperature readings. They must be accurate. Follow the manufacturers recommendations for which temperatures to take first, second, etc.

During the practice sessions, it is very important to measure and record the tire temperatures. These will tell you a lot about tire pressures and cambers. Most of the time the tire temperature readings will not be the same at the inside edge (towards the inside of the turns) as they are at the outside edge. But the middle temperature should be the average of the two side temperatures.

If the middle temperature of the tire is cooler than the average of the three tempertures, then the tire is underinflated and you need to add pressure. If it is higher than the average, reduce the pressure. Once you have found the hot pressures the tire likes, record those and maintain those hot pressures from now on.

**What Temperatures Mean** – When you look at your tire temperatures, note if a tire has more average heat than the other tire on that same side of the car. This can be an indication of a car that does not have the proper dynamic setup balance. It can also indicate a car that has corner exit problems that cause one tire to heat up from spinning.

If the RR is hotter than the RF, we tend to think the car is loose. But it may be possible that the car is tight/loose meaning it is tight in the middle and as the driver turns the steering wheel more to get the car turned. At this point, the front end quickly develops more traction and overcomes the tight condition taking the car into a loose condition.

This happens so quickly, the driver will believe the car is loose, when in fact it is tight. When you correct the tight condition to where the driver does not need to steer excessively, then the loose off condition, along with the hotter RR tire, will go away.

If the RF is hotter than the RR, it is a simple case of being very tight and you probably need to think about making spring rate changes and/or roll center and weight distribution changes. A simple way to tell if the car is tight is to note the position of the driver's hands when slowly arcing the turns. Then note again the position of the hands when the car is at speed. If the driver turns more at speed, you have a tight setup.

**Running and Evaluating the Tires** – When we run on a set of tires at race speeds, we need to look at three things to determine how well the tires are working. We can look at the tire Temperatures, tire Pressures, and tire Wear. The last two are fairly easy to read for both asphalt and dirt teams, but the first may be difficult for dirt racers to get due to the way heat bleeds off the tire after a run. None the less, if possible, get all three to help understand all that the tire is doing.

The temperatures tell us how well the tire footprint is working. Don't make major changes to your tire pressures or front cambers based on the tire temps. until you have had a chance to run the car at full speed for more than five laps. We need to get maximum temperatures in order to see the true picture of what the tires are doing.

We have learned a lot about how the tire temperatures can be an indication of ideal cambers at the front of the car. What we know now is that the ideal camber that produces the maximum tire footprint for a front tire will show the inside (towards the radius of the turn) temperature of the tire to be 15 to 20 degrees or more hotter than the outside.

If the temperatures are more even, then the cambers probably need to change. Having the correct camber is necessary in order for the tire to generate the largest foot print and give the most traction. This should be taken care of right away as we practice with the car. Do not make handling changes until the tires have the correct pressures and cambers.

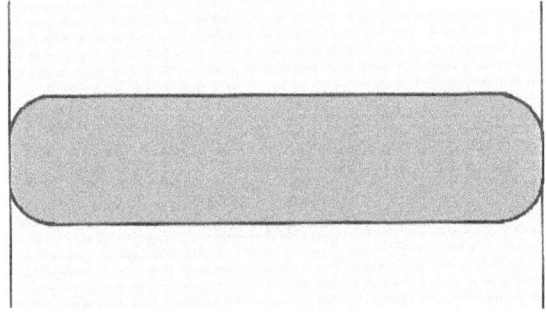

*A front tire contact patch can be maximized by adjusting the camber of the tire. We want to adjust the front cambers using tire temperature and/or wear patterns to tell if the camber can be improved. We then need to maximize the pressure distribution on the contact patch to create the largest contact patch possible. This sketch shows how our contract patch pressure distribution would look if the tire temperatures were even across the face of the tire. This is not ideal for the front tires on a circle track car.*

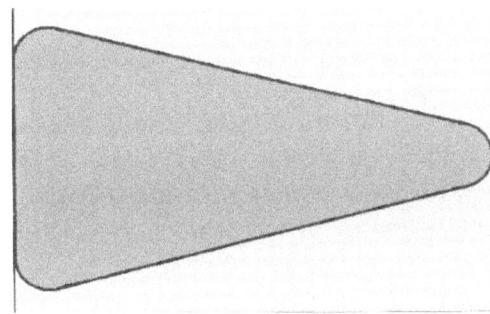

*If we introduce more camber to where the inside edge temperatures are higher by more than 15 degrees from the outside edge, and reduce the pressures, then the contract patch would look somewhat like this. The area of this contact footprint is larger than when we had less camber and uniform tire temperatures across the contact patch. Therefore, it has been proven that this cambered pattern will have more grip.*

The tire wear also tells us a lot about how a tire is working. If we measure the depth of the tire grooves or wear slots across the face of the tire, we can see if the cambers and pressures are correct. This is especially useful for situations where taking tire temperatures is not practical.

Wear pattern are especially useful for dirt applications where taking tire temperatures is not practical. We should look at a tire after having run enough laps that wear can be measured. Wear on a cambered wheel can tell us a lot about how our selected camber works with a particular setup. Also, if we are wearing out the outsides on any of the four tires, the pressures are probably too low. Excess wear on the middle of the tire tells us the pressure is too high.

Evaluating the following charts will help you see how race teams use tire temperatures and pressures to correct tire problems to create the maximum tire contact patch area.

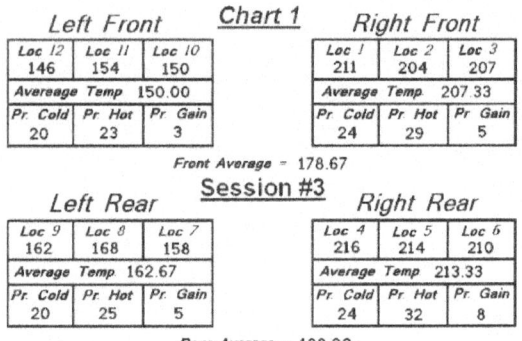

*In this chart, we have the third practice session. This is relevant because it may take several sessions to get the car up to speed and to heat the tires. Right away we see where the left side tires show a higher temperature on the middle reading. We need to reduce the pressures for the LR and LR tires. We always work with tire pressures first.*

*We see from the average tire temperatures that the RR tire is 6 degrees hotter than the RF tire. And the LF tire is a full 12 degrees cooler than the LR tire. This car exhibits characteristics of being what is termed, Tight/Loose. The LF is not doing as much work as the LR tire and this indicates a tight setup. Then as the car snaps loose on exit, the RR tire heats up causing the high RR temperatures. This car needs setup help right away.*

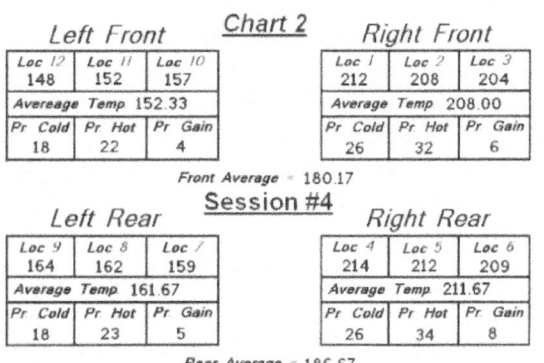

*In Session #4, the tire pressures for the left side tires have been corrected by dropping them 2 psi, but we still need to work to balance the setup so the LF tire will do more work and then gain heat to match the LR tire average temperatures.*

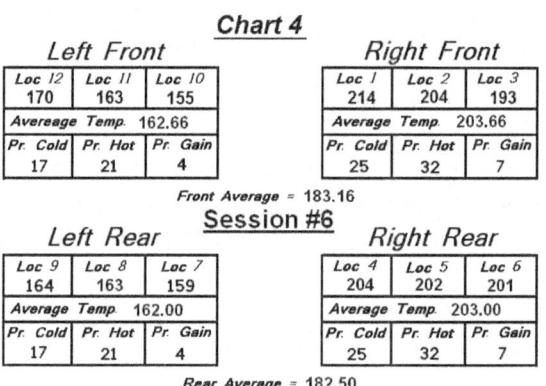

*In this Session #5, changes were made that caused more problems. The team took positive camber out of the LF tire and then also took negative camber out of the RF tire. If anything, they should have put more positive camber into the LF tire and more negative camber into the RF tire. And, nothing was done to correct the balance problem as we see that the left side tire temperatures are now still 8 degrees different. This is a tight car based on the rear averages being 6 degrees hotter than the front averages.*

### Chart 4

**Left Front**

| Loc 12 | Loc 11 | Loc 10 |
|---|---|---|
| 170 | 163 | 155 |
| Average Temp. | 162.66 | |
| Pr. Cold | Pr. Hot | Pr. Gain |
| 17 | 21 | 4 |

**Right Front**

| Loc 1 | Loc 2 | Loc 3 |
|---|---|---|
| 214 | 204 | 193 |
| Average Temp. | 203.66 | |
| Pr. Cold | Pr. Hot | Pr. Gain |
| 25 | 32 | 7 |

Front Average = 183.16

### Session #6

**Left Rear**

| Loc 9 | Loc 8 | Loc 7 |
|---|---|---|
| 164 | 163 | 159 |
| Average Temp. | 162.00 | |
| Pr. Cold | Pr. Hot | Pr. Gain |
| 17 | 21 | 4 |

**Right Rear**

| Loc 4 | Loc 5 | Loc 6 |
|---|---|---|
| 204 | 202 | 201 |
| Average Temp. | 203.00 | |
| Pr. Cold | Pr. Hot | Pr. Gain |
| 25 | 32 | 7 |

Rear Average = 182.50

*Now we're looking pretty good on the tire pressures and tire temperatures. We have corrected the front cambers to where the LF is now hotter at the inside edge and so is the RF tire. It takes this kind of temperature difference from outside to inside to produce a larger contact patch.*

*The average temperatures are now almost even on each side and the front to rear averages are almost the same as well. The tire pressures front to rear are also equal on each side of the car. This car is about ready to race.*

**Making Changes** – We want to react to any tire problems quickly, whether they are front camber issues, tire pressures or stagger. As we test and race, setup adjustments will not be correct if the tires do not have what they want to in order to provide the ideal tire contact patch.

Adjust the pressures first, and then camber as you run the car. After we are up to speed and have run several sets of laps, see can evaluate where we ended up with stagger. We are mostly concerned with the rear stagger, but make a note if a front tire grows more than normal. This may indicate a bad tire or one that is being overused. The tire temperatures will increase with the tire growth if the tire is working too hard.

Make all changes quickly after noting a problem. Correct a stagger problem right away too, so you won't chase the setup to correct a loose or tight off condition made worse by incorrect stagger.

What is obvious is that new tires are better than old ones. Tires degrade with age, even when they have never been run. Some of the chemicals that make a tire compliant will evaporate and the tires will become harder, or less compliant with age.

Don't chase a quicker car that is running fresher tires. If your lap times are down 3 or 4 tenths from a car that has newer tires on it, verses your tires that have 100 laps on them, chances are your setup is better. Once you mount your stickers (new tires), your lap times may improve 5 or 6 tenths and you will probably out qualify the other car.

**Qualifying Tips** – You need to start your qualifying laps with more pressure than if you were starting the race or going out for the first practice. There will be less time for the pressure to build when you are only running a few laps. So, a general rule is to set the left side pressures 2-3 pounds lower than race pressures and the right side pressures 4-5 pounds lower than race pressures.

In practice, many teams will put on sticker tires and run a simulated qualifying run. If you will quickly measure the pressures right after this run, you be able to see if the tires got up to race pressures and if not, make an adjustment in your notes for future starting pressures for qualifying.

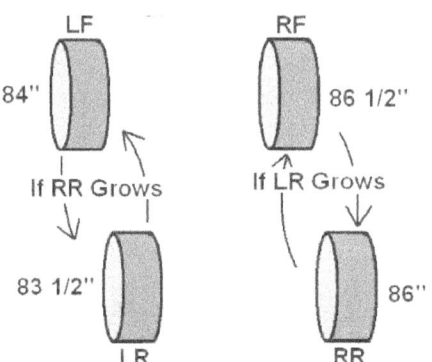

*There is a way to select and group tires so that if stagger changes are needed, you only need to swap left side or right side tires. In this example, we put the larger diameter tires on the front. If your stagger closes up, you can swap the right side tires and put the larger RF tire on the RR to bring the stagger back to where you* *intended it to be. And, if the rear stagger increases, then you can switch the larger left front tire to the left rear to reduce the stagger. The front stagger is not so critical.*

**Race Preparation** – To prepare for the race, you need to make longer runs in practice and evaluate the pressure gain. A run of 15 to 20 laps will probably tell us what we need to know. Start out with the tires cold, make the run and then take tire pressures and temperatures immediately after the car returns to the pits.

The amount that each tires grows in pressure will determine the starting pressures for the race. If you inflate the right side tires 5 pounds below your intended race pressure of say 30 psi and they come in after the long run at 32 psi, 2 pounds above our target race pressure, then we need to start out with a lower cold temperature.

The amount of difference should be a percent of the excess gain. So, this tire gained 7 pounds. The percent gain was 32 psi divided by 25 psi, or 1.28. If we divide the 30 psi desired race pressure by 1.28, we get 23.44 psi as a starting pressure, or about 23.5 psi on the pressure gauge.

For longer races, you might want to adjust the starting pressures down a pound or two. For shorter races where you anticipate a lot of cautions and shorter runs, you might want to add a pound or two to the starting pressures.

**Afterwards What to Do** – Once a set of tires has been used for practice or a race, mark how many laps are on the tires and note unusual tire growth or tire temperatures on the problem tire. If the car ran loose all night in the race, chances are the RR tire was getting very hot. Noting that may save problems if this set is used on the car for practice later on and the car begins acting funny.

Store the tires in a cool and dark place. If convenient, cover the tires to prevent exposure to the air. Used tires can always be used in practice or testing providing we understand that they will provide less performance than newer tires.

**Dirt Tire Management** - Tire management for dirt cars is somewhat more critical and more technical than for asphalt cars. The reasons why are equally complicated, but there are some basic rules to follow that will teach you where to start. The rest is just practice and evaluation. The tire guy on a dirt team must know how to read a track and what to do to the tire to maximize the grip.

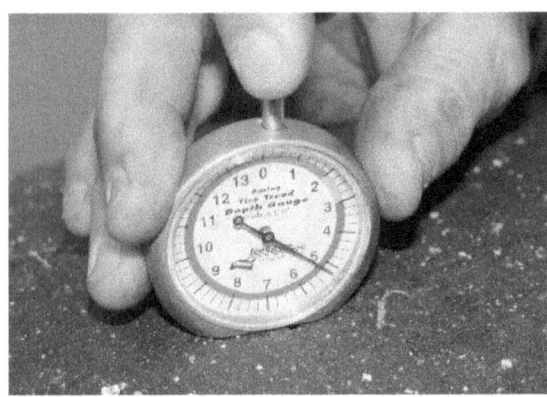

*For dirt tires, tread depth is possibly the best way to evaluate camber settings. If the inside wears excessively, then there may be too much camber. Many teams are experimenting with running more camber to create more contact patch, just like the asphalt race teams are doing.*

*In this photo taken by the author, we see a very high degree of camber in the right front tire for the number 71 car. In keeping with the theory that more camber can produce a larger contact patch, this proved this out. This car started on the pole and ran up front all day. The car to his left has much less camber and did not run as well.*

Dirt racers need to pick the right tire compound, but there is another important step to getting the most out of a dirt tire. Dirt racers often have to alter the design of a tread pattern to best work with the conditions at a particular track, at that time of the day and with their driver's style or habits. No one can tell you exactly what you will need that will work best for your situation, but there is some basic information you can use to be able to decide what to do as you encounter different racing conditions.

We have to consider the abrasion of the track, how wet or dry the track is and if the track contains rocks or other debris that will tear up the tire. All of these conditions must be considered when deciding on what kind of groove to use, how deep to cut and how many grooves can be cut out without causing the early demise of the tire. Along with these factors dirt racers must consider the amount of heat a track puts in their tires.

In addition to enhancing traction, grooving helps the tire dissipate heat and can be used to help control tire temperatures.

**Grooving and Siping For Dirt Cars** - Tires used on dirt cars will undergo some special preparations prior ever hitting the track. The one preparation process is grooving tires. The main job of a grooving tool is to place a greater number of cutting edges into the tire to aid in gaining grip from the dirt surfaces it is racing on and to help cool the tire.

*This is an electric tire grooving and siping tool that uses heat to help cut through the rubber. The tire is grooved so that there will be more edges in the tread and more grooves help cool the tire. There will be accelerated wear when grooving and/or siping, but the team can judge the wear factor when evaluating how much to groove the tire.*

There are three basic shapes used in grooving: 1) Square, 2) V grooves, and 3) Sipe cuts. Square grooves are the same width through its entire depth. V grooves start out wide at the top and taper to nothing at their bottom. Sipes are thin slits cut by installing the blade upside down in the holder and using the separate ends of the blade to cut slices in the tire.

All three types of grooves can be used in various depths depending on the conditions and the length of the race. The V groove is often used when the track is expected to need more tread contact later in the race. As the tire wears, the grooves become smaller or disappear completely. Square grooves can be used the same way but the extra width at the bottom of the groove could provide enough leverage on an abrasive track to tear the tire if the track becomes tighter.

Siping is usually meant to make the tread more pliable and does not produce the edges the Square or V grooves do. Siping also helps the tread maintain a more consistent wear that helps keep the tire working uniformly.

Grooving the shoulders can be helpful if you plan to run the high line or on a cushion and need to be moving some dirt. Grooves cut into the shoulders help clean away some of the loose dirt to get at moisture beneath it. Sometimes the outer row of blocks is also grooved on the right rear to work with the shoulders.

On some of the harder, natural rubber tires teams sometimes sipe the shoulders. This can really help when you are rolling the tire under when running lower tire pressures on a very slick track. The sipes can help prevent the shoulder area of the tire from glazing over and losing traction. The shoulder is as important a part of the tire as anything else and if you are going to be running on it due to low air pressure or because you are running against the cushion you need to make the best use of it.

The angle at which grooves are cut determines how much of the edges are exposed to the track when the car is in various degrees of slide. Dirt race cars seldom if ever run in a straight line. When the track is tight we keep the grooves pretty straight because the car is being driven more straight ahead. But when we spend a lot of time with the car sideways on a slick track, we put more angle into the grooves.

How much of an angle is dependent on the driver's style and experience is the only way to determine the best angle for your situation. The idea is to keep the maximum amount of the tires edges facing the direction the tire is actually traveling.

The grooving process is something that any racer can learn and there are a good number of tools on the market that make it a very simple process. They range from a simple sharp razor blade type of tool to heated knives. The heated knives, or grooving irons, are the logical choice for this process. They are faster and they give you a greater level of control.

Most dirt racers are using blades in the range from .125, .187 and .250 of an inch in width. The power required to run most grooving irons is no greater than what would be required to run a 3/8" electric drill motor. So, you will not have to use a large capacity generator to run a grooving iron at the track.

Any time a groove or sipe is cut into a tire it accelerates wear. The trick is to balance the benefit of grooving with the increased wear. One of the reasons we sipe a tire is to prevent the tread surface from glazing over and becoming slick. Sipes help the tire to wear and to keep it working throughout the race.

Learning to recognize the amount of wear you can expect from a track not only helps you choose the best compound tire for the night but is important information for deciding what pattern and depth of grooving to use. The object is to maintain the highest level of traction throughout the race without using up all of the tread with five laps to go.

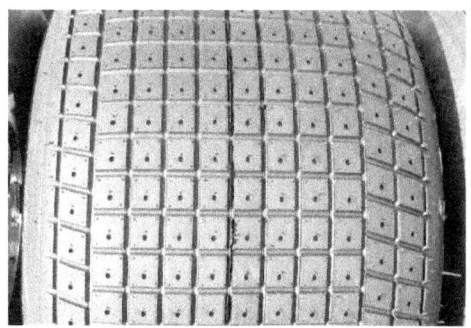

*This is a new Hoosier dirt tire before any grooving or siping has been done. Note the square blocks, These blocks will be cut into smaller squares by grooving if needed. Those smaller squares can be cut by siping to create more edges in addition to the grooves.*

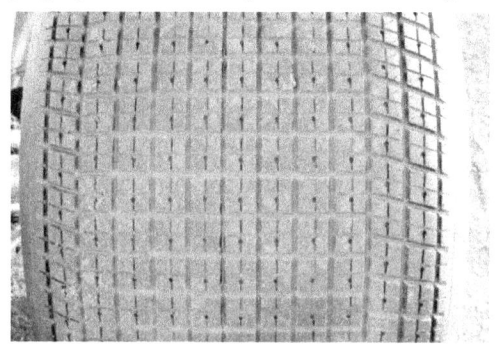

*These blocks have been grooved and cut into four smaller blocks. This tire has been used and exhibits a lot of wear to the point that the grooves cut into the tire are almost gone. Note also that the middle 2/3 of the tire is worn, but the outer parts are almost like new and still have the molding remnants. This tire could have possibly been run with less air pressure.*

*This is a closeup of a tire that has been recently grooved. Note the cut across each of these two blocks. These cuts are beneficial for a slick track and will provide more edges to grip the track plus provide more cooling.*

*The angles of the grooves can be any angle relative to the direction of the tires path. They can be straight across like those in the left middle section as opposed to the right third with diagonal grooves.*

**Note Keeping** - The first step in proper tire management is to keep detailed notes. This is your first line of information gathering. You can purchase a multitude of pre-printed setup sheets from a variety of racing vendors. In fact, many of them are free. The issue is not so much the format but the fact that you are keeping notes.

You most likely will develop a format that works for you over time, but in the beginning you will need to keep track of a few things from a tire management perspective that may include but are not limited to:

1) Brand and compound number of each tire. All tires have a S/N, or a manufacturing date, or other unique identifier molded into the tire that will help you single them out

2) Tire age, meaning the number of laps the tires have on them. You should be able to document the number of laps each tire in your inventory may have accumulated to date.

3) Which corner the tire was run on,

4) Tire and wheel width size,

5) Air pressures (cold and hot),

6) Circumference of the tire, pre and post run,

7) Tire grooving for dirt tires (size of grove, pattern and depth),

8) General condition of the tire pre and post race, as in are the edges sharp, is the tire feathered or blistered, and

9) Tire temperatures recorded after each run.

Creating a note sheet with each of the parameters you are tracking in a "fill in the blank" type of format will make this data much easier to keep track of and much quicker to do. The easier and quicker it is to develop or execute, the more likely you can get the crew to accomplish the task.

Make this task as easy as possible. It will become important to designate a person on your team who always has this responsibility, or you can rotate this job between several team members from race to race to further develop your team. These note pages need to go into a master notebook that you will be able to refer to for future setup information.

When you have developed a good note keeping routine, you can refer to the data and it can help you to understand just how the tires are performing. It also helps you to keep a better level of control over the number of tires you keep in your inventory. You will know which ones to sell and which ones to keep and know just when new tires are required.

**Search For Information** - Asking questions or searching online are ways we can glean information. Your first resource is to go to the manufacturer's website for information, compare this information to your notes and use this information to further improve your individual race tired data bank.

The information would include, required optimum wheel widths for a given tire and the recommended maximum air pressure settings. Find out what the recommended tire temps are for the best performance and tire life. On the same thought learn what temps are too high and inversely which temps are too low. The majority of the tire manufactures have web sites that explain and answer all of these questions.

Goodyear and Hoosier both have great web sites for the dissemination of this type of information. Your next resource should be the dealers who are selling you tires. They want you to do well with their tires, that way you will buy more as the need arises.

**Summary** – Like we have stated, all that we do with the setup of the car is for the benefit of the tires. Tires provide the final ingredient for performance. Tire loading and tire contact patch optimization equals high performance for any race car. Race tire optimization is track dependent in many cases. What works for one track may not work for another.

A lot has been learned in the recent past about how to optimize the tire contact patch area. It is known that a tire with a larger contact patch, with the same loading, will provide more grip. The shape of the contact patch itself will ultimately determine the amount of grip.

To learn more about how to manage your tires at your track, or with your series, look at what the other teams around you are doing. Develop a rapport with your fellow racers so that you can ask and learn. Share information you might have learned in order to gain their trust. And talk to the tire pros at the track and at the factory.

# Exam - In The Context Of This Lesson:

## The Most Useful Tools Used For Working With Race Tires Include?

1) A large dial pressure gauge
2) A narrow and flat tape measure
3) Nitrogen gas
4) All of the above

## When We Take Tire Temperatures, We?

1) Record the surface temperature
2) Check the temperature at the middle of the tire
3) Probe deep into the tire at three points
4) Can use an infrared digital thermometer

## Race Tires Are Measured How?

1) Around the middle of the tread
2) At normal racing pressures
3) With a narrow and flat tape measure
4) All of the above

## Race Tire Stagger Is When?

1) The cambers are different side to side
2) When the front tire circumference is larger than at the rear
3) The inside tire circumference is smaller than the outside circumference.
4) All of the above

## If The Middle Of The Tire Is Cooler Than The Average Tire Temperature, Then?

1) The pressure is too high
2) There is not enough camber
3) The pressure is too low
4) There is too much camber

## If The Rear Tire Averages Are Hotter Than The Front Tires?

1) The setup is tight
2) The car might be tight/loose
3) The car is loose
4) 2 and 3

## Dirt Tire Grooves And Sipes Are Cut Into The Tire For?

1) Increasing the number of edges for more grip on dry slick tracks
2) To cool the tire
3) To better clean the tread
4) All of the above

## Cutting V-grooves Into Dirt Tires Is Useful When?

1) On tight tracks that will go slick
2) On slick tracks that will go tighter
3) For dry slick dusty tracks
4) For tacky tracks

## Lesson Twenty – Summation RCT Level Two

Congratulations, you have completed the Race Car Technology course, Level Two. This final Lesson will remind you about all that we have learned in this course. We outline all that we have accomplished in taking you to the race car. In Race Car Technology Level Two we have taken all of those parts and pieces we described in Level One and applied them to an actual race car. And, we presented all of that important information in a logical order, just the way you would approach an actual race car and prepare it for competition.

*Before we do anything to setup our race car, we first check out all of the parts to make sure they are able to hold up to the rigors of racing. We told you how to go through the entire car to make sure it is track ready and safe.*

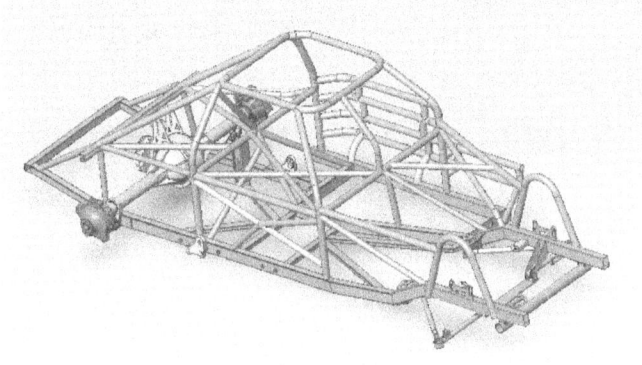

*After learning all about the parts and pieces of the race car, we now, in RCT Level Three work on the actual race car using all of those parts. We go from one end to the other carefully explaining how to setup and use all of those systems.*

We first outlined all of the goals of Level Two. We proceeded to go through all of the systems on a race car and help you understand how to work with the race car to accomplish the stated goals.

We ten taught you how to go through the car to make sure it is track worthy. We taught you how to set your ride heights, distribute your weights and what those weights should be.

In the geometry sections, we discussed the settings and considerations, caster, camber, bump steer, Ackermann and alignment. In the alignment sections, we went over front and rear alignment, rear steer, toe and the driveline alignment.

Once our car is aligned correctly, we moved on to the spring selection in and how to arrange our spring rates in order to achieve dynamic balance. As a part of the springs choices, the sway bars play a part and act as springs. Then we described how we use shocks to control those springs.

*We then set the ride heights and established the corner weights we will need to go racing.*

In advanced setups, we described how to setup the race car using bump technology in todays racing. Bumps stops and bump springs are being used in almost every form of motor racing today.

The next item we think about for our setups is the Anti-dive and Anti-squat settings. We fully explained how those are designed into the suspension and when, and when not, to use the Anti's.

*We went through the settings for caster, camber, bump steer, etc. to get the chassis setup for the type of racing we will be doing with each specific type of race car.*

*The alignment of the wheels is the next step we covered and explained in detail how to properly align a race car for each type.*

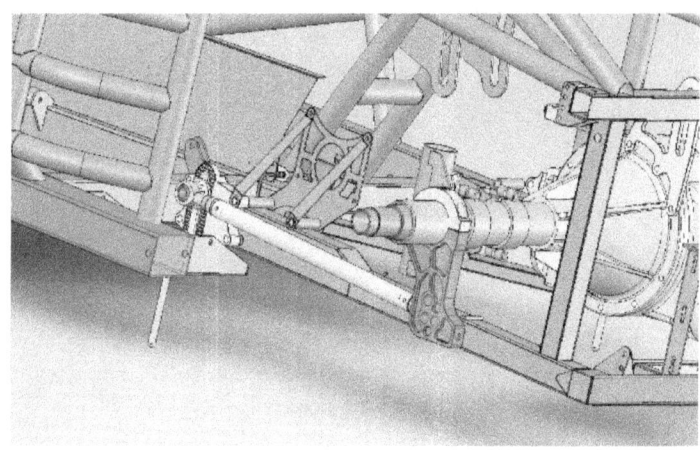

*Next came the explanation of how to use rear steer to enhance your drive off the corner and other uses for an advanced understanding of the technology of rear steer.*

*We told you how to check Ackermann and when it is, and when it is not, appropriate, and in what amounts.*

*You were able to gain valuable knowledge about spring selection, roll stiffness and roll angle dynamics as well as how to control those springs using shocks.*

*Another current trend and one that is used in almost every form of modern racing is the bump devices. The bump stop, or bump springs, add to the spring rate of the ride spring and we told you how that works and how to adjust your shock package to work with the bump setups.*

Now we have to go to the track and actually test and race our race car. We outlined how to read and select your race tires. After all, the tire is the final point of contact between your race car and the race track. All that we do involves making those four tires work as hard as they possibly can.

*The final thing to go over before we actually take our race car to the track is tires. We describe how to choose and mount the tires. We also tell you how to read the tire temperatures and pressures so that you can get the most grip out of your racing tires.*

And once we get all of the systems set correctly and we have our tires mounted, we are ready to test all of that. We tell you how to conduct your test sessions to get the most out of the race car. We outline what to adjust for what is happening with the car, in an order that makes sense.

*When we have done all that we need to do to setup our race car, the day comes when we take the car to the race track for the first time. We tell you how to weigh the car and how to conduct a test session to get the most out of the experience.*

Now that you have all of this information, you should be well on your way to setting up your own race car. In RCT Level Three, we continue with our Lessons by describing advanced setup technology and current routines and processes that have never been presented before, anywhere.

*Now it is time to run the car. We outline what to evaluate once the car is on-track and how to make changes to the setup for dynamic balance and proper load distribution. We tell you how you can make your car faster and how you will know when it is working properly.*

Level Three actually goes on to explain the engineering behind setting up real race cars. We take you through the process in detail describing what we are doing and how exactly to do it on an actual race car. There are methods and knowledge that has only recently been discovered and applied to race cars, and in Level Three, we explain and present all of that knowledge. We look forward to your continued education in race car technology in the ORS continuing education series of courses.

*Hopefully with all of this newfound knowledge, your team can extract the most performance out of your race car that is possible. Then maybe you'll be able to visit the winners circle and take home a trophy.*

# Exam - In The Context Of This Lesson:

### The Primary Goal For Setup In A Race Car?

1) Make it easier to drive
2) Increase grip
3) Balance the two suspension systems
4) Create less chassis roll
5) 2 and 3

### Why Should I Go Over The Whole Car If It Was Running When I Bought It?

1) To make sure it is correct for the rules you race with
2) Because it might have the wrong parts
3) It is your responsibility to make sure it is safe
4) It probably does not have the correct setup to be competitive
5) All of the above

### We Maintain A Defined Ride Height Because?

1) The suspension angles will always be consistent
2) The static loading on the tires will stay consistent
3) Aero downforce will be predictable
4) All of the above

### The Following Can Change The Weight Distribution?

1) Sway bar pre-load
2) Front end travel
3) Travel onto the bump stops or springs
4) Camber Changes
5) All of the above

### We Can Change The MC Location By?

1) Raising or lowering the chassis mounts
2) Using taller or shorter ball joints
3) Changing the lower control arm angles
4) Changing the upper control arm angles
5) All of the above

### Caster Settings Are Relative To?

1) The track banking angle
2) The track turn radius
3) The type of race car
4) Driver preference
5) All of the above

**Why Do We Make Camber Changes?**

1) To make the car turn better

2) To create a better heat pattern on the tire

3) To create a larger contact patch

4) All of the above

**Ackermann Is Present When?**

1) The wheel steers when it bumps

2) Only one wheel turns

3) We gain or lose toe when the wheels are turned

4) Our steering arms are too short

**One Way To Create Power Rear Steer Is?**

1) Using the squat of the rear from weight transfer

2) Using the acceleration force driving the car forward

3) Using the torque of the motor

4) All of the above

**Proper Driveline Angles Are When?**

1) The pinion is angled 2 degrees down to the front

2) The transmission shaft and pinion shaft are parallel

3) The drive shaft angles at the transmission and pinion are equal and opposite

4) 2 and 3

**Matched Roll Angles Create What?**

1) Ideal weight distribution

2) A faster turn speed

3) More overall grip

4) Better tire wear

5) All of the above

**Shocks Control What?**

1) Speed of suspension movement

2) Suspension forces

3) The rate of the spring

4) Transitional handling

5) All of the above

**Bump Springs Have The Following Characteristic That Bump Stops Don't?**

1) A consistent spring rate

2) A wide range of spring rates

3) More spring travel before coil-bind

4) All of the above

**Race Tires Are Measured How?**

1) Around the middle of the tread

2) At normal racing pressures

3) With a narrow and flat tape measure

4) All of the above

**The First Part Of The Track We Tune For Is?**

1) Turn entry

2) Mid-turn

3) Turn exit

4) Straightaway speeds

# About the Author

Bob Bolles has been a hands-on motorsports engineer for over twenty-five years. Although he holds a B.S degree in the Mechanical Engineering, his skill and experience with racecars comes from working directly on the chassis and with many hundreds of race teams. He likes to think that he comes from much the same mold as a few others before him such as his friend, the late Dr. Smokey Yunick, who also wasn't afraid to "get his hands dirty" to effect change. Like Smokey, Bob loves this sport and enjoys the interaction with others who also love it.

The nice thing about racing is that you cannot BS what you know or what you develop. It either works or it doesn't. Proof is just a few laps away. Before Bob started his research into racecar dynamics and engineering, he struggled with the very same problems that most race teams continue to struggle with today. He just knew there was a better way and if he just looked hard enough, the answers might come. Well, they did. That information is shared in the pages of the RCT series.

Over the years, Bob has worked with a high degree of success on virtually every type of stock car raced in the United States, as well as sophisticated formula type race cars. Whether it is an asphalt stock car, modified, or a dirt late model, his techniques and methods have improved handling and opened the door for winning. He has engineered cars that have won major asphalt late model championships, touring championship, modified championships, road racing championships, and dirt late model races including the Dream and the World 100 at Eldora Speedway.

It is this broad level of experience and his development of new technologies that qualifies him to write about these important subjects. Not a day goes by that Bob isn't speaking with or helping racers. With contacts all across America, he is truly in tune with the pulse of auto racing on a technological level that very few individuals enjoy.

The software Bob developed in the mid-1990's is still being used by championship winning teams throughout the U.S., Canada, Australia, New Zealand and Europe. His dream of being able to help all racers in their pursuit of success and enjoyment is fast becoming a reality.

Bob has been a professional motorsports technical writer serving as the Senior Technical Editor for Circle Track magazine for over fifteen years and currently as a technical contributor to Speedway Illustrated magazine.

Race Car Technology- Level Two represents Bob's second book in this series. Bob is also the author of RCT - Level One and RCT – Level Three. These books are being used by college instructors to educate future motorsports engineers.

www.ingramcontent.com/pod-product-compliance
Lightning Source LLC
Chambersburg PA
CBHW081828300426
44116CB00014B/2513